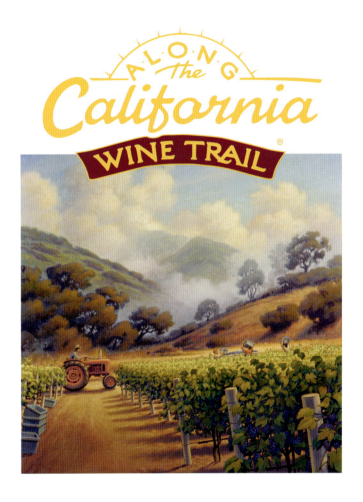

Along the California Wine Trail

"Do not go where the path may lead, go instead where there is no path and leave a trail."

—Ralph Waldo Emerson

WRITTEN BY **JERRY STROUD**
PAINTINGS BY **KERNE ERICKSON**

GREG YOUNG PUBLISHING, INC.
PO Box 2487
SANTA BARBARA, CALIFORNIA 93120
gregyoungpublishing.com
greg@gregyoungpublishing.com

Copyright © 2017, 2018, 2019 Greg Young Publishing, Inc. for all Kerne Erickson artwork
with the exception of the Santa Catalina Island painting, page 102:
Copyright © 2011 Geoff and Alison Rusack

Copyright © 2019 Jerry Stroud for all written story content
"Type Font"– Patent No. D812679, Greg Young Publishing, Inc.

All rights reserved. No part of this book may be reproduced or transmitted in any form or by
any means; electronic, mechanical, or digital, including photocopying.

UPDATED EDITION

Written by Jerry Stroud
e-mail: script434@aol.com • www.alongthecaliforniawinetrailauthor.com

Artwork by Kerne Erickson

Design by Jennifer Pechette • Layout production by In Color, Santa Barbara

Printed in China by Integrated Communications • www.icla.com

ISBN: 978-0-578-60826-6
Library of Congress Control Number: 2019955234

Cover Artwork by Kerne Erickson
Front: Kunde Family Winery "Mountain Top," Sonoma Valley
Back: Fiddletown

To purchase this book or artwork, please visit:
californiawinetrail.com

Table of Contents

Chapters

FOREWORD 4

PUBLISHER'S NOTE 6

ACKNOWLEDGMENTS 7

INTRODUCTION 8

SIERRA FOOTHILLS 11

THE NORTH COAST 28

THE CENTRAL VALLEY . . . 58

THE CENTRAL COAST . . . 70

THE SOUTH COAST 93

Paintings

Alexander Valley 42
Amador 25
Anderson Valley 54
Arroyo Grande Valley 80
Arroyo Seco 76
Ballard Canyon 83
California Shenandoah Valley 13
Calistoga 45
Carmel Valley 72
Clear Lake 50
Dry Creek 29
Edna Valley 78
El Dorado 21
Fair Play 22
Fiddletown 16
Fort Ross Seaview 39
Happy Canyon of Santa Barbara 88
Lodi 65
Los Olivos District 85
Madera 59
Malibu Coast 99
Mendocino 55
Monterey 70
Napa Valley 47
Oakville 44
Paso Robles 74
Paso Robles Geneseo District 79
Russian River Valley 28
Rutherford 46
Santa Catalina Island 102
Santa Lucia Highlands 71
Santa Maria Valley 84
Santa Ynez Valley 87
Sierra Foothills 11
Sonoma Valley 34
Stags Leap District 48
Sta. Rita Hills 89
St. Helena 49
Temecula Valley 94

Foreword

Every bottle of wine tells a story, and behind those stories, there are often generations of families whose lives are shaped by the magic of the grape and the challenge to create something spectacular. This is true for me, as it is for many of the wine country pioneers you'll read about in this book.

I first fell in love with Paso Robles while I was at UC Davis in the Enology program. There were 17 of us. At that time, Napa and Sonoma were already far ahead in the industry of making premium wines. Paso Robles was in its infancy (Mine became the fifth winery there) but I fell in love. My professors promised me this would be a great wine region. I believed them and I wasn't alone. Efforts from people like Herman Schwartz, Vic Roberts, Tom Martin, and Jerry Lohr also helped put Paso, and Paso wines, on the map.

I remember those early years being filled with a lot of hard work and a lot of fun. But my story is not unique. With every wine appellation in California, there are pioneers who pave the way. Even though today's wine industry is far different from the one I started out with, we're all making great wines but facing different challenges, technologies, and regulations that we must work with. What matters most is a love of the land, the soul of a winemaker, and the perseverance to succeed.

When I started out in 1973, we were blazing a trail because we knew in our hearts that California would become one of the finest wine regions in the world. New generations may choose to follow our course or forge their own path in this industry. Either way, I can attest that California creates some of the best wine in the world. All you have to do is open a bottle to find out.

Cheers,

Gary Eberle

Publisher's Note

My journey down this trail began in early 2011. Winemaker Geoff Rusack and I were having a discussion regarding the Ballard Canyon area in the Santa Ynez Valley of Santa Barbara County. Geoff was defining Ballard Canyon's unique climate and topography, which produces world-class Rhône varietals. While we discussed this, I had a thought: how can I assist Geoff with creating more awareness of Ballard Canyon — a very unique and beautiful place, that is not often traveled and few people are aware of? It was during this conversation I started thinking of a concept to accomplish this. The initial idea of promoting awareness to one specific wine-producing region quickly grew into a statewide project. I soon realized my endeavor to promote California's American Viticultural Areas (AVAs) would be no easy task, since there are well over one hundred in California. I thought to myself, what better way of introducing people to California's AVAs than a story told by Writer Jerry Stroud and paintings by Artist Kerne Erickson? The journey down this trail was met with enthusiasm from everyone I encountered.

I'd like to thank Jonathan Walters, Chair of Lake County Wine Grape Growers Commission and Director of Farming for Brassfield Estate Winery, for his time, sharing of his great knowledge, and the suggestion that we experience the marvelous sunrise over Clear Lake. We spent that entire day touring and exploring Clear Lake sub-AVAs; I will always remember and appreciate the experience and Jonathan's natural enthusiasm.

I am truly appreciative and grateful for the creative efforts of writer Jerry Stroud and Editor/Designer Jennifer Pechette, whose endless patience and hard work made this book a reality. And to Gordon McClelland: thank you for introducing Kerne Erickson, whose paintings are like windows, providing the viewer a taste of the awesome bounty that the wine trail has to offer. Thank you Kerne, for your creativity, endless hours, and dedication to the project.

Thank you to my dear wife Ashley and my children, Jozef and Mia, for their support, patience, and love. They are the Bright Stars in my life.

There are so many other individuals who made invaluable contributions, I cannot express my gratitude enough. Thank you! The completion of this book — this joyful and wonderful feat — could not have happened without each and every one of your efforts.

Greg Young

President, Greg Young Publishing, Inc.
Santa Barbara, California

Acknowledgments

The immediate attraction to author *Along the California Wine Trail* was based upon several ideals. California was home. Wine was a shared joy, and to express words that complement Kerne's nostalgic paintings was a pairing that resonated with me. The concept was fresh, something untold. A wise man once said, "Opportunity has one hair on its head. When it passes by, you've gotta grab it."

So, I did.

And while I did, I put on hold the lost Oceanic art of a man whose story will be told another day. The decision was based upon a desire to pursue a journey that was, at its very core, my own. It was a trail I had lived, whose ranchers and farmers I had grown up with, whose countryside I had wandered through. To share stories of the lives who have helped transform this complex world of winemaking, in places I still call home, became deeply personal.

And who more passionate a man to champion California's viticultural story than Greg Young? It was Greg's vision — plus his ability to communicate and inspire with crushing perseverance — that was paramount to the book coming to fruition. The growers' stories and vineyard images would not have graced these pages if not for this gentleman publisher.

A sincere thanks goes to the book's editor and layout designer, Jennifer Pechette, whose page-turning notes were both invaluable and inspiring. Her layout structure and fresh design add a creative element to the book that beautifully complements the words and paintings throughout these pages.

Thank you to my beautiful wife, Suzanne, whose patience at home was unending, and whose companionship on the road allowed us more time together as we experienced California's delicious rewards along the way.

My thanks go out to Kerne Erickson. Having the privilege to share words alongside his buttery-soft and alluring paintings is a pairing more than I could have hoped for.

With that said, my sincerest thanks go out to those of you who shared your time and stories with me as I traveled the open road. Your lives touched me in ways that I did not expect. So, as the book reflects California's beautiful sense of place with nuances of color and expression, so does it reflect the best of individuals themselves. To each of you, I'm truly grateful.

Jerry Stroud
Author

Introduction

Being a native Californian who has lived from one end of the state to the other has helped influence my feelings about this leggy-shaped place I've called home for nearly 60 years. Having worked as a cellar rat to assistant winemaker, I've come to know a bit about California and the wines produced by it.

But, I am neither a wine critic or a wine-speak snob. I consider myself as much qualified with wine speak as Robert Parker Jr. is with critiquing toast. I do know wine delights my senses. It brings friends and family closer, enhances those joyous moments seated around the table, and makes every occasion feel as if there is a sense of holiday spirit to it.

What is drawn from the age-old drink comes from more than the mere must of grapes. The life given to it begins with terroir. It is sense of place that matters. The winemaker finesses and sculpts character, but a wine without place would be like an actor without a script. Without a story, nothing can be told. This is the significance of place. It is in her bedrock subtext; her sun belt bench lands; her shrouding fog and ocean breezes; her rainstorms and sun-drenching open skies; and the rock and soil beneath her.

To best understand her character — that personality so influenced by place — is to have lived here one's entire life. There is no better complement to a story than experience and no better influence than having explored and understood those relevant matters. It is to know California from the inside out; to have a personal understanding of her intimacies, her breath and her smile; and like a good friend, to have shared in her triumphs, and been beside her in tragedy. These are quintessential elements to a writing companion.

In my youth, I spent Sonoma Valley evenings couched on the strapping arms of colossal oak trees as a full moon crested over the straw and volcanic outcroppings of unplowed hills. I have felt the fury of her earthquakes and witnessed the tragedies of her firestorms. I have hiked through private Sonoma Mountain outbacks and have trekked the Sierras to greet the most magnificent ponderosas. In my early 20s, I cared more about hunting a legendary wild hog named Volkswagen in the Frei Brothers Winery foothills than tasting the wines those brothers had to offer. I preferred to walk the majestic land now covered by Warm Springs Dam waters to boating it. To this day my heart goes out to several redwood patriarchs buried beneath those waters.

I have fished for bass in Lytton Springs ponds; hunted dove in Alexander Valley along the Russian River banks steeped with gravel; jump shot migrating duck in Geyserville; tromped through rice fields after Canadian Geese in the Sacramento Valley; and cooked trout in the high Sierras. I took Bodega Bay party boats out to the Cordell Banks in the stormiest seas. I caught ling cod and red snapper and buried their carcasses beneath my Santa Rosa garden. I have taken my boat to Catalina Island, anchored in coves, and caught lobster and Bonita. I was even married on the island at Mt. Ada. I have tasted firsthand the elements of the Pacific Ocean waters, breathed the piney air, and know well the sweetness of the whipped fog topping coastal ranges. I have worked the bottling line and in wine cellars, painted the red iconic barn at Sonoma Vineyards, mixed in my share of bentonite, and stirred my share of lees. I have been seduced by the purest free-run juice from chilled tanks and have, with athletic delight, dribbled empty French oak barrels from one end of the cellar to the next — and within those cellars climbed more ladders than the corporate elite.

But it was in my youth that, like those buttery soft notes so common in California Chardonnay, the beauty of the state sang a tune to me. It wasn't the wine, but the allure of the land. The sense of place called and mountain influence mattered. So, in the late 1970s, I quit Sonoma Vineyards and left home and family (virtually everything) to explore what I believed was the most rugged part of northern California. I wanted to know the state's other side. I knew of her beauty, but I wanted to be challenged by her strength. I longed to spend a season in the Marble Mountain Primitive Area with everything I owned on my back: to be free of it all, to feel her earth beneath me, and to be humbled by her majestic mountain tops.

The adventure validated the allure and infatuation I had with this sense of place, so magnificent and far removed from the Sonoma I knew so well. I returned to work for a small winery named St. Francis Vineyards. It was under the ownership of Joe Martin during its boutique days. From cellar rat to assistant winemaker, I suddenly embraced a winegrowing region that resonated with me, beneath the shadow of Sugarloaf Mountain in Sonoma Valley. Life was good. The all-night harvesting, racking, and pump overs, along with meeting winegrowers and ranchers from the industry was a privilege. However, none were as impressive as a warm and gentle character I'd been introduced to named Andre Tchelistcheff.

It was Andre's overwhelming presence in the cellar, his patriarchal charm and commanding respect that helped me, a young man by mere acquaintance, to better understand the relationship between man and place.

Along the California Wine Trail is a toast to this place called California and a gift of profound thanks and appreciation to those lands I grew up with. It's a handshake to the ranchers and many winegrowers who allowed me access to places I would never have known and to wines I would otherwise never have had the chance to share. More importantly, it is a tribute to those individuals, both past and present, whose tireless efforts remind us everyday of humanity's strength, of life's generosities, and its delicious rewards.

Jerry Stroud
Author

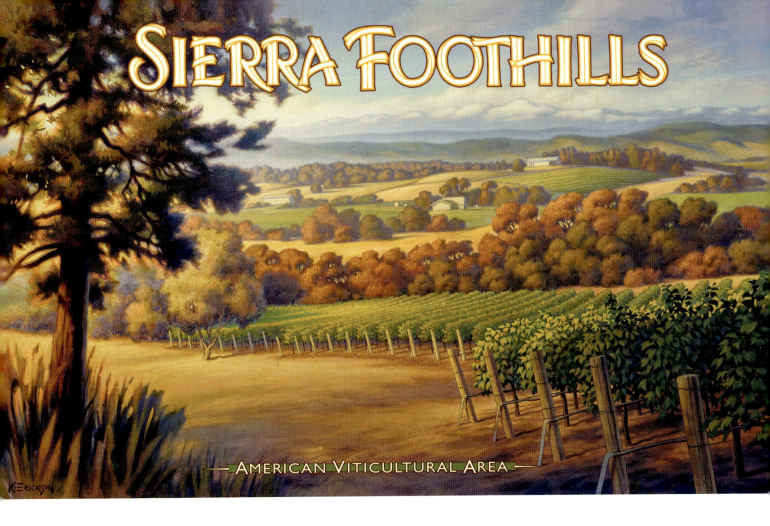

1

Sierra Foothills
American Viticultural Area

"They say we're gonna have an inch of rain tomorrow," Dick Cooper, owner of Cooper Vineyards said as we drove slowly by a block of Barbera, his Labrador Blondie by his side. Dick looked toward the vineyard with his signature cowboy hat shading his brow. I didn't quite know if he was talking to me or to the vines passing by, reassuring them that it's gonna be okay.

This is how ruthlessly dry the season had been in California. His remarks also reflected how he, and many other Foothills grape growers, are connected to their vineyards. His appreciation and attention to their temperament and personality was as genuine as his smile. He can't hide the fact that this ever-growing relationship between man and vineyard is born by family ties — generations of them — and whether tapestried in October colors or bud breaking naked in spring, they speak to him.

Dick is not alone in his respect for the Sierra Foothills vineyards, or belief in those wines born from them. In speaking with others from the Shenandoah Valley, like owner Andrew Friedlander of Andis Wines, there is a movement going on. Not only is the Shenandoah Valley producing award winning-wines with their signature Zinfandel and bold Barbera, but planted acreage throughout the region is at an all-time high. Grape growers are coming to the Foothills because place matters and they are able to introduce new varieties favorable to the region.

It is a renaissance defined by change, leading to a broader swath of Rhône variety plantings. Grapes like Roussanne, Mourvèdre, Grenache, and Marsanne are thriving. Well known varieties like Syrah and Viognier are bringing character to blended reds, adding diversity to wines for those enthusiasts seeking something more. Established grapes like Sangiovese, Barbera, and Primitivo are increasing in production, along with more signature varieties predominant to the Foothills like Zinfandel and Cabernet Sauvignon. The Foothills are on the move! And despite the recent seasons of drought, the vintages have been impressive and the region continues its march forward.

Since the Foothills' resurrection in the '70s, planted acreage has increased by the thousands; yet with this ascent, the region has not reached the pinnacle of respect it deserves.

The five sub-appellations within the Sierra Foothills AVA — California Shenandoah Valley, El Dorado, Fair Play, Fiddletown, and North Yuba — have been slow to mature. There has been little notice of the region's viticultural significance; some refer to the Sierra Foothills as having sub-par viticulture. In Andrew Friedlander's words regarding their assessment, "It's been tough." Despite this, the vision for the future is unwavering, and speaks of the growers and vintners' commitment and belief in what lies ahead. They have tastings, winemaker dinners, and vintners association marketing efforts. They host wine festivals like Amador County's Big Crush, in celebration of the grape harvest, with nearly 5,000 attendees, all of this toasting the region's success. Down the road from Andis Winery in the Shenandoah Valley is another winery where the owner believes in the region's future. His name is Dave Helwig.

Relaxing on the cushy chairs, Dave Helwig shared his story of the winery's modest beginnings, based on the philosophy, "If it's meant to be, it will happen."

The foothills are on the move!

Dave began his first venture in grape growing and home winemaking in the Southern California enclave of Thousand Oaks, where he planted 50 vines and half a dozen varieties. The self-proclaimed Midwesterner realized this wasn't enough. So, he and his wife Nancy decided to go large.

They moved from the Pacific Coast and purchased a beautiful piece of land where the Foothills meet the sky. Here, they founded Helwig Winery. Today the winery is producing award-winning wines and the Helwigs have estab-

lished themselves as a major player in the Shenandoah Valley.

Dave's satisfaction with their endeavor was apparent when he reflected on evenings he and Nancy shared while gazing across the winery grounds. "Can you believe this is ours?" he recalled asking. Dave didn't expand on those shared moments, although it was clear something profound happened in their lives. It wasn't about them being owners. It was more about a couple's vision. There was a sense of fulfillment to the new addition in their lives.

Giving birth to Helwig Winery happened in 2009 when they began their first crush. Back then, their son Scott Helwig was the cellar rat; today he is the winemaker. Scott's influence on wine is achieving attention; the 2014 Marsanne and the Rosé de Shenandoah were both awarded double Best of Class at the 2016 California State Fair. And sure to steal the show is a new member of the Helwig's family of wines of distinction: a collection of premium wines respectfully dubbed DAVANCY, for Dave and Nancy.

The grapes contributing to these award-winning bottles come mainly from Helwig's own vineyards. They source few grapes. With 60 acres they have no need. Only neighboring Barbera grapes from Dick Cooper's vineyards travel to Helwig from across the road.

Like many tasting rooms built above valleys that touch the sky, Helwig was born to impress. On a typical day you can see the Sierra Mountains, and on an exceptionally clear day, the Pacific Coast Mountains.

Dave speaks fondly of clearing land for grapes. The trees removed from the property were beautifully repurposed. Oak and ponderosa were milled and stylishly incorporated in the winery's tasting room. The large oak slabs are now bar tops and the interior walls are lined with ponderosa. The decomposed granite soils excavated from wine caves are packed throughout outdoor walkways. The entire winery is indigenously crafted with Shenandoah Valley's unique sense of place.

Just below the walkways and outdoor lounge is Helwig's quaint amphitheater in the sky. Past concerts headlined '70s era bands like Pablo Cruise, Pure Prairie League, and Dave Mason: bands that were once at the top of the page.

With all that is happening in Shenandoah Valley, it still holds an idyllic charm to its presence. While some residents fear the region's growth will change that, others feel growth can be responsibly managed. When I asked Dave what challenges the region faced in terms of growth, he was clear. "A lack of lodging," he said. Apparently there are more ponds and open pasture for Canadian geese looking for room than there are rooms for visitors seeking refuge. Today, the majority of lodging is located in Folsom, Eldorado Hills, and Sacramento, all about an hour away.

> The winds offer a whispering of fragrances … notes of pine needle, blooming buckeye, and tannin scented oak.

Ultimately, this region will experience growth, but Dave isn't worried about its future. "If it's meant to be, it will happen."

The altitude and character of the Sierra Foothills speak of a higher calling. The soil is consistently composed of volcanic loam above 1,500 feet, with sandy loam of predominately decomposed granite below this level. The terracotta-colored iron topsoil is visible at almost every cut bank. The depth of the soil varies from several inches to several feet and is considered by many viticulturists as being suitable for Rhône varieties. Growers from across the country are coming to this immense AVA with great aspirations as there are thousands of acres of valleys, rolling hills, and steep mountainsides suitable for growing a host of varieties. The Sierra Foothills AVA covers eight counties and encompasses 2.6 million acres, approximately 5,800 of which are planted with vineyards. More than 125 wineries operate in the area. Those looking to bring new character to the bottle are turning to this affordable, virtually untapped region.

The terroir can best be described by Jeff Meyers, general manager at Terra d'Oro, who refers to the Sierra Foothills region as having a Mediterranean climate. Like Spain and Italy, a vast number of its microclimates are consistently higher, sunnier, and steeper than other wine regions in the state. The Foothills begin at 500 feet and climb to 3,000 feet toward a clear sky, rising above the noise and industrial lees of pollution that often settle in valley floors. The Foothills reside closest to the grand Sierra Nevada mountain range to the east. These mountains form a wall that stretches north to south for approximately 400 miles. From Fredonyer Pass to Tehachapi Pass, Pacific storms have stacked up on its western side and blasted the face of the Sierra Nevadas for eons. It was exposure to these countless storms that eventually decayed the mountains to form the up-and-down character, vast array of soils, and microclimates that make up today's Sierra Foothills.

Here, the region draws breath from down-drafting alpine winds and warm upward winds from the valley floor. The variety of trees that thrive in the high-altitude sun are natural born filters. The winds offer a whispering of fragrances: notes of pine needle, blooming buckeye, and oak. The Foothills have a consistent climate with mostly warm days and cool nights. The region may encounter several days of fog and a dusting or two of snow per year, but it is the morning frost at bud break when the vineyards are most vulnerable. The damning chill, with its dagger-like follicles, can smother a vineyard in minutes. In the unprotected lands throughout Amador County, you will see wind machines rising above vineyards. These towering sentinels stand guard over the vines, helping to raise temperatures in an effort to protect the vines' tender shoots.

Later that morning, Dick Cooper pointed out some irregularities in a straw field beyond the vineyards. "See that?" he asked. "It's from early miners." The miners certainly traveled far and worked hard — Dick respects that — and I didn't press the grievances I have regarding the early miners. Unlike many, my own understanding of the era is a period in California history that pilfered the very sense of place that is at the heart of the vineyard — the land.

It's easy to understand the lore and why the legacy continues. It's easy to fall into romantic trappings as dreams of riches called to the vast number of immigrants who traveled here. It's also easy, as a self-proclaimed historian, to see how their transgressions were not only systematically sugar-coated, but muddied with falsehoods.

During the late 1800s, many unsuccessful gold prospectors turned to grape growing and wine-making, but other gold-seeking immigrants continued to burrow beneath the vineyards. The hydraulic techniques used by miners were a clear and blatant threat. The farmers were at arms with them, and due to this atrocity, a law was passed in 1884 banning hydraulic mining, which protected farmers from the invasive wrath ravaging the earth beneath their feet.

More importantly, Sierra Foothills farmers were practicing sustainability before they knew the significance attached to it. They were the Uptons, Uhlingers, Deavers, and the D'Agostinis, whose families respectfully pioneered Shenandoah Valley in the mid-1800s, as well as the Sobons, who were handed the torch from the D'Agostinis in 1989 and have carried the light of responsible farming into the era of sustainable practice.

Today, they are professional growers like Andrew Friedlander, whom I had the pleasure of meeting. He's an educated man whose "green" Andis Wines are a mark of environmental and architectural excellence. They are vintners who encourage visitors from across the country to view the area, educating them on the importance of what's happening here. They are families inviting guests to share in the abundance of Foothills wines, pouring glasses in tasting rooms throughout the region. They welcome the traveler onto their property. In fact, you will be hard-pressed to find a more amicable, approachable region than the Sierra Foothills.

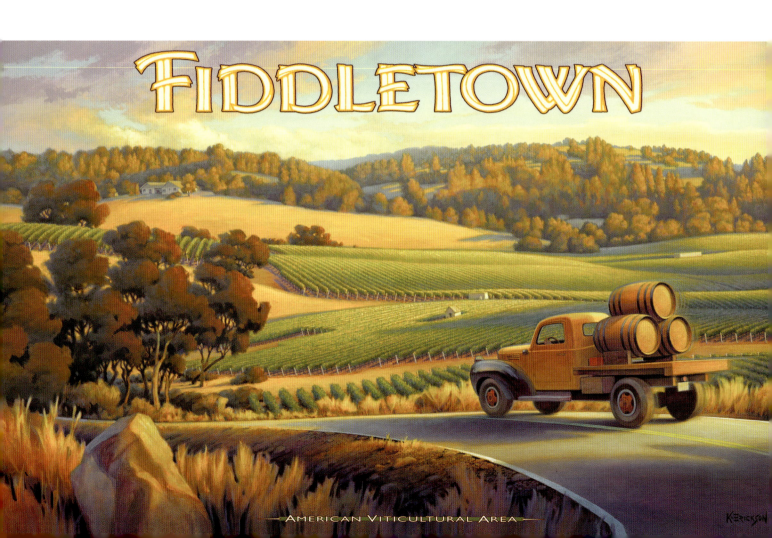

FIDDLETOWN
AMERICAN VITICULTURAL AREA

There is the new generation of winegrowers as well, transcending the formidable past with the promise of tomorrow. This progress is not made due to any sense of snobbery or elitism. It is made by a breathing, living coexistence with the land, the animals, and the communities with which we share our world.

History tells us the miners brought the first vines to the Foothills in the Coarse Gold Gulch area. That was in the mid-1800s and is historically significant. These were early settlers — immigrants who planted vineyards and established cottage wineries. The Possessory Act in 1852 allowed the Foothills farmers to file for ownership and patent acres of land for farming, land they would eventually call home. This encouraged farming and was good reason to begin the endeavor of planting vines. And plant they did; the Sierra Foothills is said to have been the most prolific wine growing region in the state during this era. By the late 1800s the region was saturated with as many as 100 wineries during its crescendo.

By building houses and planting vineyards, the farmers discovered their own sense of place here, one that would last for generations. They planted more than vines; they planted family roots in the land. They understood the importance of environment and the significance of conservation. It is what established the wine growers' place in the Foothills: a fit between land, grape variety, and people. This sense of oneness inevitably led to the wine industry's growth. It was not thoughts of nuggets dancing in their heads. They came with a plan that decades later inevitably led to the greatest boom: the winegrowing bump in the 1970s. Crop demand skyrocketed, and the value of Foothills grapes were recognized for their intensity, spice, and fruit-packed boldness.

Foothills grapes were recognized for their intensity, spice, and fruit packed boldness.

It was the Great War, followed by the Volstead Act in 1919, the collapse of the mines in the 1920s, the Great Depression, the stock market crash, and eventually the Second World War that led the wine industry to be punished by circumstance. The economic wrath had produced the perfect storm. And even though Prohibition was repealed in 1933, it wasn't grapes that were sought out for planting, but walnut and prune orchards. Grape interest in the region was literally crushed, pressed back decades by uncontrollable events due in part to the Gold Rush collapse.

Yet in today's Sierra Foothills, an unavoidable association with wine and gold continues. Jingles like "the hills are filled with grape gold," or "the new gold rush of the Sierras", or "liquid gold," are often repeated. The old clichés continue throughout the region. I understand the legacy.

The bread and butter of the wine industry's past begins with its own history. There is no place to rip a chunk of bread from its remarkable past than at Sobon Estate, whose story speaks of tireless preceding generations made of responsible individuals prospering in communion with the land and each other.

The Winery's collection of artifacts contains old vats, tools, and equipment that once pressed the fruit and worked the fields. Images of pioneering men who worked the land adorn the walls. These are the remarkable stories of responsible individuals who closed deals with handshakes.

I'm a history guy. California is my home and mining is its legacy. It will forever remain an important piece of California's architecture. But to continue chirping about the miners' association with wine is a colossal mispairing. The blend sours the taste buds and waters my eyes just thinking of it. In fact, if I hear another "Strike it rich in the Sierra Foothills" in reference to wine, or the "New Gold Frontier" as a simile for the region by local rags, brochures, or newspapers — that's it! I'll drive straight to Jackson, find a saloon, and get hammered on some cheap southern whiskey.

There are blogs talking about the Foothills providing wine for thirsty miners during the gold rush, and there is no question they did, but the term "thirsty" for the miner in relation to wine is a stretch. I'm not talking about toothless immigrants in cranky saloons sniffing Chardonnay from GraniteWare mugs, grumbling about buttery notes and burping up sourdough biscuits. Most of these guys worked in and around water all day. It came with the job. So they had plenty of time to quench their thirst and mend their pruning toes. But quenching thirst with wine? The majority of miners were a hardcore bunch of drinkers whose incessant need for alcohol and all-night benders certainly wasn't quenched by wine as much as it was by whiskey and other hardcore spirits that packed a snort and swallow. The best bang for the coin came with a drink that taunted law and tortured livers.

I understand the gold rush jingle provides traffic to hotels, shops, and the general community. Traffic means dollars and dollars make sense. The gold miner's legacy continues just as the hotels of the era continue to thrive with their stacked brick walls, creaky porch walkways, tin overhangs and antique consignment shops enticing travelers like myself to step in. They should. It's a rich, genuine story ideally built for tourism as well as a great snapshot to California's rowdy past. It's Foothills history. I get it, but the Sierra Foothills are producing award-winning wines. Today's pioneering individuals are expanding the region's footprint beyond the wine clubs, local markets, and tasting rooms. Travelers to the Foothills are raising their glasses to the region's success and filling their receptors with the complexity of an aromatic pairing. The alpine and ponderosa-scented air paired with a fruity bold Barbera resonates with this delicious sense of place. At stellar tasting rooms with stunning vistas, the wine enthusiast can indulge in what is arguably some of the finest Zinfandel in the state.

This is the allure of the Sierra Foothill AVA: its wine. The Foothills are more than just gold country with a sidekick. Wine country is the story here, and the promise is what's happening today. The region's continued diversity in planted grapes is transforming life near the mountain tops. But the silence here is deafening and is one of the challenges facing the Sierra Foothills. There's need for noise.

Sure, the tranquility here is sublime and its charm unforgettable. The feeling of escapism is part of its attraction and unquestionably a source of its viticultural appeal, but the Foothills need to be heard. People need to get here and the region needs to be seen. Yet its footprint takes us off the four-lane freeway. OMG! It leads us away from the cherished Apple stores, pop culture coffee houses, and beyond that "fifty miles and I won't eat sushi" threshold. These are conveniences the pampered public bathe in. Even locals will tell you the drive home is a "poke" from any major city—out of the way for some, too far for others.

The Foothills need to be heard. People need to get here...

However, the road traveled today is changing. It is one of the many reasons the Foothills are growing. Attention to highways and country roads have transformed immensely, and with today's automobiles — their safety, mileage improvements, and technological advancements — is there really any inconvenience? I thought my parents '60 Ford station wagon was space-age stuff, with its ultra-cool rear power window. Today we have navigation, satellite radio, and a host of other advancements embedded in our cars. It doesn't matter if you're snaking along the coast highway or doing the ups and downs between Napa and Sonoma or Lake County; today's vehicle allows the traveler unparalleled riding comfort, even on the most challenging roads. They are living rooms on wheels; space-age, technologically-driven devices that steer their own wheels, for God's sake. What else does the pampered traveler need? When driving from the valley floor to the Foothills, it's as calming as if I'd just turned onto the last road home. My greatest challenge is fighting off a nap. It's that comfortable.

As I leave the mass traffic through Central Valley, pass Lodi, and traverse the first incline heading towards the Sierra Foothills, I take a breath, immediately knowing the craziness is behind me. I'm closer now to my destination — that place away from the busyness city life stacks on me. I find the road up here simple and calming. The drive is so visibly appealing I'd pay to travel it. In fact, having lived in Orange County, I'm surprised there are no toll roads charging me to get there. It's that beautiful.

So, if you're looking for a memorable drive this is it. More importantly, if you're a wine enthusiast seeking a higher calling, there is no better time than now to pop a cork, experience the delicious character of Foothills wines, and make some noise!

Dick Cooper also showed me a healthy block of Barbera and said, "My father told me, 'Son, don't ever plant here. It's too gravelly,'" he smiled in recollection. Then he chuckled respectfully, pleased at the sight of the healthy Barbera vines that overcame the odds — not gloating over success in the face of his father's advice, but proud of the vines flourishing in the face of diversity. It is this determination — that gut feeling farmers like Dick have for their crops — that leads to a closer connection to the vineyards. In many ways it's paternal — a rare instinct born by sheer affection for the vineyard and experience with the land. It isn't science. It isn't book talk. It's less of a business to the farmer than it is a fulfilling lifestyle. It's years of a life dedicated to caring for this sense of place and for the vines who've found their home here.

From north Yuba County to Mariposa County, Foothills farmers like Dick Cooper have a vision of sustainable farming. A new green is happening. It is more than talk; sustainable, natural, organic, and biodynamic practices have been adopted by the grape grower. Many grape growers have their own philosophies regarding these practices. Dick

Cooper doesn't subscribe to organic farming. He believes in sustainability and like many, understands the community's own sense of place within the farm. The $5,000 of bird netting he purchased is rolled up in the barn. He prefers planting rows of bird feeders along chosen spots to keep the migrating birds from ravaging the grapes. The predators are doing well with rodent control. He uses his farmed sheep manure for fertilizer. He dry farms and cultivates alternating rows. There are as many variations in farming practices today as there are wines to choose from.

I asked Jeff Meyers of Terra d'Oro about biodynamic farming. "Yeah, I know about it," he said. He, like many, regard it without much fanfare. The reality is, biodynamic is only a fraction of today's farming practices. Wineries are reluctant to cross over, usually because they don't believe in biodynamics, or because transforming their farming practices to meet the strict requirements can be extremely challenging.

It has a paramount risk for any grape grower, especially the established. It's not like adopting sustainable practices, where you address cover crops, go solar, change to biodiesel, or fabricate bird houses for predators to control rodents. With biodynamic farming, you're picking crops by the stars, pruning vineyards by the planets, and burning mice pelts by the dozens. That's right, actually torching them to be sprayed into the air upon the vineyard. Sure, these things are at the far end of the practice spectrum, but to be Demeter Certified, these standards do apply. It is changing a winery's entire growing philosophy; implement something as ideological as a palm reader into the house and leave everything you've invested to the wine gods. And the truth is, wineries like Terra d'Oro have worked tirelessly to get to where they are. Things are working in the Foothills.

Certainly, tradition is meant to be challenged. Like Dick Cooper, whose respect for his father's beliefs were successfully challenged in the vineyards, it should come as no revelation that change is imminent. Farming practices will inevitably transcend today's methods as they will transcend tomorrow's; that is the beauty of the industry. This isn't a tale about new, but the story of old. It's been evolving for centuries and will continue to do so at the behest of another generation seeking non-traditional methods themselves. The demand is happening today.

The wines produced in the Foothills have many renditions to their story. Classic traditional wines, like the Foothills' signature Zinfandels, are similar to a Shakespeare story; all have a good number of interpretations with nuance, personality, and human influence distinctly their own. Whether these wine stories are characterized by terroir or cellar influence, some will continue to please us, others not.

Among the vineyard blocks and tilled rows, a growing number of other stories are waiting patiently to be told. They may pale in comparison to their peers, but they are scattered among the Foothills acres, waiting for the vintners' hand to bring them some measure of prominence. They are the "other wines," yearning to be understood and talked about for who they are. The millennial generation will be the decider of tomorrow's wines. They will champion those grapes that bring another character to the bottle.

As every new wine tells its own story, so does the Foothills winemaker whose influence in the cellar creates a personal rendition of them. Whether it's cold fermentation, organic practices, or new French versus American oak barrels, the winemaker's craft can complement or break the story.

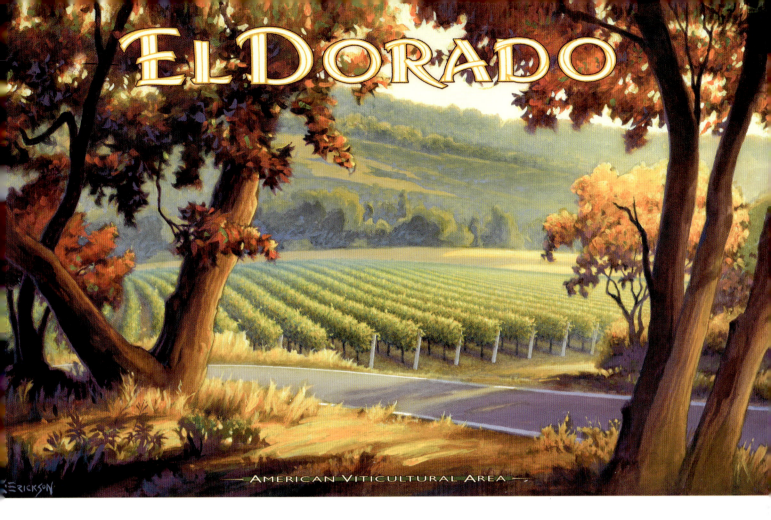

North Yuba is the smallest sub-appellation in the Foothills. It consists of 30 square miles with less than a dozen wineries in its region, while its neighboring El Dorado sub-appellation has over 60 wineries. Within the southern area of the El Dorado AVA sits the Fair Play AVA. Close to 20 wineries are located among its twisting turns and mountaintop terroir.

One example providing a loftier view of Fair Play is Skinner Winery. From the tasting room upper deck, I was impressed with the heavy snow blanketing the Sierra Mountain range during my summer visit — a rare experience when traveling California's immense vineyard wonderland.

Here, you can enjoy a noon tasting and later that day ski the vista you were just viewing. Or, spend the entire day toasting the good life, then whisk yourselves off to your favorite Tahoe resort, cozy up by the fire, and enjoy heavenly ski slopes by morning.

> One of Fair Play's greatest qualities is the absence of crowded tourist trappings.

Unlike the much-chirped about American Viticultural Areas where coveted dirt, blessed vineyards, and celebrity talent reign over the land, Fair Play's boutique wineries offer a sense of place with far less fanfare. Select wineries set out roadside placards reading "Great Food," and local musicians play gigs now and then. While tasting, it's not unusual to meet an owner or winemaker as they pass through in rubber boots, stained hands,

Along the California Wine Trail — 21

and pomace dangling from their cheek hairs. They come out of the darkness of cellars or sun-drenched vineyards to share a lifestyle — one that allows them time to breathe and develop character without the hurry.

> **What the farmer "sees" comes with age. It comes with experience and is drawn from the countless struggles and triumphs born by them.**

One of Fair Play's greatest qualities is the absence of crowded tourist trappings. You may not be afforded sip-and-see tram rides, nor will you float high above vineyards in woven baskets tethered to colorful balloons, but you also won't be dishing out extravagant costs "glam wineries" have imperially corked into the price of their bottle.

A fair question is, is the trip worth it? Does the reasonable pricing reflect something less in the wines, or are these simply undiscovered gems?

Maybe. Nobody really knows, unless they've been there, and that's one of the less-talked about nuances of on-the-road tasting. It's really about discovery. Let's face it, wine tasting isn't simply about taking snobbery notes and paying attention to the spit bucket, unless you're "one of them." It's about seeking out wines that delight our senses. For many, it's also about the joy of exploration — stopping in between wineries by that rickety old barn on the roadside and snapping that selfie, or discovering a personal connection without plugging into one.

At the heart of the Sierra Foothills, in the county of Amador, lies two other sub-appellations that complete the Sierra Foothills AVA: the California Shenandoah Valley AVA and its adjacent Fiddletown AVA. Here you will find over 50 wineries and a plethora of other unique experiences, one of these being the refreshing enclave in Plymouth called Amador 360 Winery Collective. Though it has no scenic view, its "360-degree view" promise of lesser-known Amador wines is an experience worth tasting. You'll be introduced to wines produced by unknown makers, part-time viticulturists, as well as established winemakers with their own Foothills brands. Just around the corner you will find Taste restaurant which features an exquisite menu, a fine wine list, and a cast of professionals that could've came straight from a five-star metropolitan restaurant. It is the go-to place for local and out-of-town wine professionals.

The local wines talked about are many. Some prominent, others not. They consist of a cornucopia of varieties, enough Old World names to tongue tie even the most educated Rosetta Stone enthusiast. Over 30 varieties — some indigenous to places like Spain to Italy — are grown throughout the appellations. Albariño, Vermentino, Fiano, Verdejo, and Greco Nero flourish. White wine grapes from France's Rhône Valley, like Roussanne, Marsanne, and Viognier are also present.

These high-alcohol wines, referred to as "hedonistic fruit bombs" by Robert Parker, Jr. are prominent in the Foothills. They will continue to be championed, while a movement towards Rhône and Burgundy varieties of a more elegant note is concurrently changing the viticultural landscape in the Foothills. As people tire of a boring sameness that's common in today's wine landscape, diversity will overcome homogeny. This is the Sierra Foothills' sweet spot. It will not come with

Along the California Wine Trail — 23

a smash, bang, or immediate celebration. It will come gradually — day by day, year by year — and this movement will eventually transform the industry. It will be led by today's generation, and the generations that follow– individuals who are not only in search of their own identity but who seek to define themselves and their pleasures by those elements enhancing their daily lives.

Later that day Dick said, "I'm sorry about kidnapping you for so long. I just like driving around my vineyards. I like looking at them." It was a telling statement. Like a great wine, there was complexity to what influenced those words. "Looking at," defined by the grape grower, is not something visible to those of us unfamiliar with this paternal connection they share. What the farmer "sees" comes with age. It comes with experience and is drawn from the countless struggles and triumphs. In Dick Cooper's case, three generations of them.

As we puttered between the vineyards at a tractor's pace in the comfort of his truck and best friend Blondie, I could see Dick was in his element. He was able to eloquently express the deep respect he has for these vineyards that consume him. By explaining how his wines came to be and how he's cared for them throughout the years has helped him define himself. Just hearing him speak of their company seemed to soothe him. His own words validated generations that have looked over this place. The putter through the vineyards was a giving of thanks to family and to the land he so loved. It was no wonder he "just" liked looking at them. They spoke to him. It was as if he was somewhere else — and in far better company than with me — when he spoke of them. The more he reminisced, the further back he traveled. The years came seamlessly to him. He was now in a time prior to the vineyards when his grandfather farmed hay — fuel for horses who powered the carriages.

While these early days working the hayfields were met with great challenges, the farmers were able to overcome them. It was this longevity and wisdom that earned them a respectable place in the Foothills. To this day, there is no bellyaching, no whining, no resentment brought on by hardship, only solutions to them. Some solutions work. Some don't. The lines drawn in their leathered skin and solid handshakes reflect their toil and resolve. It's that simple. They continue working, tirelessly seeking better ways and knowing well the importance of tomorrow.

The fact that some of the oldest Zinfandel vines in the world come from the Sierra Foothills is reason to talk about them. Those gnarly vines — with their flaking skins and strapping arms — are testaments to the region's viability and rootstock grit Vineyards, estimated to be planted as early as the 1840s in the Coarse Gold Gulch area, appear to flourish with the wears of age. They go back generations, many vineyards outliving the very lives that brought them here – a compliment to every grower.

Jeff Meyers believes there's a good possibility all Foothills Zinfandel clones derive from the original Deaver clone. There is a profound sense of celebration of "old" throughout the Foothills region as well as the wine industry itself. While ageism is an issue throughout much of corporate America, here in the wine industry, old is embraced! At the very heart of old age abounds a true sense of spirit. It is genuine. Old denotes a silver lining. Think of the old farmer, the old vine vineyards, and the old wine. Old is as valued and refreshing as a bud-breaking spring. When one views the old

AMADOR

vines among the Foothills and tastes the character in their grapes, its reverence is understood.

Today, the Grandpère Vineyard in the Shenandoah Valley is a fitting example. The 143-year-old vines, planted by John and Mary Upton on a gentle slope of Foothills land, continue to produce high quality wines. Equally impressive are the crops harvested by Terra d'Oro from the Deaver Vineyards' 140-year-old Zinfandel vines. These are ancient hipster vines with decades of torrential rainstorm parties and ground shaking earthquakes beneath their roots. They are rock stars, inciting envy in other regions. The word "old," when referencing vineyards, not only has a sense of celebrity attachment to it, but a soulful spirit associated with the vineyards as well as the wine. The label sells bottles. It is in many ways an agricultural darling. Where else in today's society will old experience that kind of love?

When asked about what threshold legally determines old in regard to the vines, Jeff Meyers just laughed. "Three years old can be considered old." How's that for a slippery slope? I have iPhones older than that.

Yet, these 143-year-old ancestor vines on this 14-acre piece of Amador County land is testament to the Foothills triumph. They speak of wisdom. They are the region's legacy and their clonal variations will become ancestor vines themselves.

The following morning a rainstorm blew in to the Sierra Foothills, just as Dick Cooper had promised. It crept in early to the pleasure of many. I

Along the California Wine Trail — 25

stepped onto the historic Imperial Hotel's balcony to look over the grey wet past. If history had a face this was it: Amador City, the heart of generations born and gone. Old was preserved everywhere. The tin roofs, porch overhangs, and clunky boardwalks resonated with dirt from the past.

Looking over Main Street I understood the importance of California's history. It was at the heart of the town's legacy, and yet, the Sierra Foothills has respectfully outgrown its historic bygone years. More fittingly, the region is now a slice of wine country's future. Like a farmer's old pair of jeans, it is the perfect fit.

While the morning rains continued to drench the city's hardscape, I was reminded how the region deeply needed a good drenching. Looking over this antiquated town, I thought of those pioneering grape growers who had to care for their parched and thirsty vineyards without technological advancements. Their struggles brought on by drought, pestilence, and other hardships were unending. But history tells us the vineyards thrived. There were no drip systems, no sprinklers, no pagers beeping vineyard managers that frost was imminent. There were no pads to tap or apps to assist, no weather Doppler to inform them, and more importantly, no dead batteries to charge.

There was only the Farmers' Almanac. It was the go-to reference book for every farmer. Nothing more. The fact is, it is still Mother Earth who decides what the day will bring. Ultimately the sun, climate, and soil are what determine life among the Foothills and the outcome of every vintage.

This is the farmer's sustenance as well — Mother Earth. It is what they depend on. Without her benevolence, all of mankind's crops would wither and render even the best technological advancements passé.

Grape growers know the importance of the environment. Their awareness is nurturing a better understanding of those things tied to earth. They are wrapping their minds around responsible farming practices, and by doing so, are cultivating the importance of tomorrow. It was good to see the Sierra Foothills vineyards in the care of these responsible farmers — generations of them — whose confidence in the region's future is to this day unwavering.

While leaving Amador wine country, I felt like I was driving away from a dear old friend. I had learned something more than is simply taught or seen among the vineyards; there was a sense of place here. The fresh drafting alpine air, with its soft fragrances of oak, pine and buckeye, delivered without condition. To share in the company born by these natural elements was to be entranced by a place you'll find nowhere else. Is it any wonder the word Amador means "one who loves?"

As I drove home the rain continued to fall. It was good to see the fields drinking and the gullies swelling. The vineyards, bathed by the rain, seemed to have a sense of celebration to them. Harvest was in, and even though it had come two weeks early, it surely pleased the farmer, who had cause to celebrate. I envisioned growers and vintners raising glasses at dinner tables across the

Sierra Foothills, giving thanks to this gloomy, wet, and grey presence. It wasn't much, but it was a start; helping water tables to rise, assisting leaves to color and fall, and allowing vineyards, exhausted by the years of challenging drought, to rest in preparation for the coming year.

I thought of that sun-drenched morning with Dick Cooper, his elbow hanging out the open window, not knowing if he was talking to me or to the vines, but knowing this: there was more to his words than a promise that rain would come.

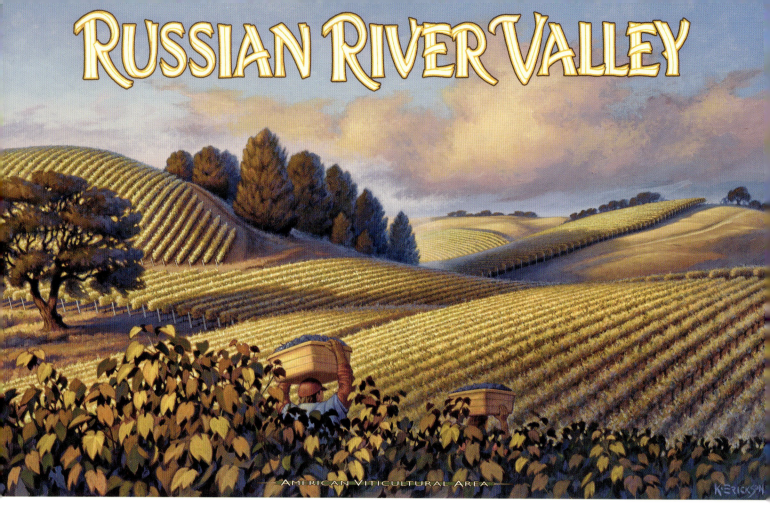

2

The North Coast
American Viticultural Area

"If you drive through Napa, you pass castles. Drive through the Russian River Valley and you pass barns." This was Francine "Frankie" Williams, the owner of Toad Hollow Vineyards, expressing with a sense of pride the wine regions' cultural differences. It's clear: The Russian River Valley today resembles a place I could have easily driven through decades ago. Its charm is as timeless and valued by those who live here as the faux castles and grand chateaux are to those who reside along Napa Valley's Silverado Trail.

The Russian River Valley is one of many sub-appellations within the North Coast AVA. It is an acclaimed growing region whose wines like Pinot Noir and Chardonnay thrive in its coveted soils.

A short walk from Frankie's vineyards, along Westside road, is a block of vineyard that looks as plain as any other — rolling hills in the backdrop, oak groves, straw-blonde hillsides — yet this is no common block of vineyard, but one that helped bring something spectacular to a bottle and revolutionized the wine industry in California. It's a vineyard that helped define the North Coast region wine as the best in the world.

This is Bacigalupi Vineyards, whose contribution of grapes to Chateau Montelena helped win the West. It took a blind tasting across the pond for the French — whose wines have for centuries defined best in class — to realize that the New World, and specifically Napa County's Chateau Montelena's 1973 Chardonnay, would outclass their own to become legendary.

Titled "The Judgment of Paris," *Time* magazine's foreign correspondent George Taber broke the news of the May 24, 1976 results. Taber's four-paragraph story was at the time considered filler; today it remains a story that helped changed the course of California wine history.

The Burgundian terroir so chirped about by French oenophiles was suddenly second class. British wine merchant Steven Spurrier attempted to bring California wine to light by pitting New World wines against French wines in a blind tasting. Surprisingly, the white wine that triumphed was from Napa's Chateau Montelena, a 1973 Chardonnay produced by master wine-maker Miljenko Grgich.

It was a seminal moment in wine history. But if the grapes contributed to the Napa Chateau Montelena wine really didn't come from Napa, should it matter? What if the distinct soil, the microclimate, and the elevation that help to define character and personality in this particular vintage had

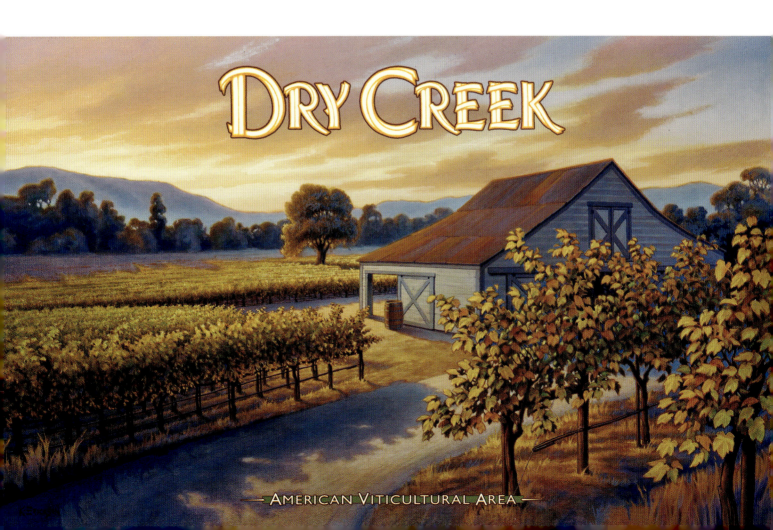

Dry Creek

— AMERICAN VITICULTURAL AREA —

come from another appellation? Several, in fact. What if, in reality, a percentage of the grapes had been sourced mountains away from the Napa Valley in a viticultural area designated as the Russian River Valley, in a modest block of vineyard owned and managed by Charles and Helen Bacigalupi?

I had the pleasure of sitting down with the 90-year-old matriarch, Helen Bacigalupi, on a foggy morning at the winery's tasting room. Her husband Charles had passed away in 2013. Beside Helen was her granddaughter Nicole, who along with her parents and twin sister Katherine, has taken over the reins to the Bacigalupi Vineyards.

Helen, a sharp-witted, business-savvy woman, recalled the day Mike Grgich called her from Chateau Montelena saying, "'Did you hear? Your grapes were used in the Chardonnay that won the Judgment.'" There was a proud sense of accomplishment to her smile. I asked Helen if the Paris Judgement changed the course of her and Charles' winery as it had for those others involved in its success? If it brought acclaim to them? "Not so much," said Helen. "It did though, make Mike Grgich rich," and "put California on the map." She laughs, shrugging off the respect gained by all those involved with the Judgement but her. After all, they were only grapes. They were only a large percentage of two other Sonoma County grapes used in the making of the 1973 Napa-labeled vintage.

After my interview, Lee Hodo, a local wine marketing expert with 36 years of experience, and currently representing the Bacigalupi family, drove us to the block of vineyards near Helen's home that contributed to the seminal event. The vineyards appeared unremarkable.

I picked up some vineyard soil and ran it between my fingers. The earth was warm and crumbly. Harvest had come early. The pregnant bunches bulging from the vines had been delivered weeks earlier and the remnant grapes left to hang were now measurably sweetened by summer's end. Giving birth was reflected throughout this block of seemingly forgotten vineyard. Days of green had passed, the vines appeared exhausted. It was time for a season of rest. The dormant sleeping days of fall and winter were ahead, until another year of warming begins the vines' journey, a season of growing that one day might give birth to another spectacular vintage to be remembered.

Before leaving the Russian River Valley, I had the privilege of lunching with Frankie Williams at Wily's in Healdsburg. Frankie, a Blythe Danner look-alike, expressed her joy and passion for the business after having taken over Toad Hollow Vineyards in 2007. Her late husband, Todd, had passed away unexpectedly, leaving her to carry on a legacy that continues today.

Frankie shared with me some candid thoughts on the industry: the business is a "jungle" out there. The people are "a joy." The competition is "huge."

When I asked Frankie what she was most proud of, there was no hesitation. It was her ability to carry on with the winery and to continue what Todd had left behind — a desire to provide fine wine at affordable prices.

She said a wine critic once remarked, "Frankie, you'll never get the respect your wines deserve at these prices." She laughed. There wasn't disappointment or regret. In fact, for Frankie, there was a sense of satisfaction in what concerned the critic. That was the mission Todd had championed since inception — affordable prices! This approachable and much-appreciated journey continues today; classy, delicious, whimsically-branded wines that more people can enjoy.

This approachable and much appreciated journey continues today— classy, delicious, whimsically branded wines that more people can enjoy.

All too often experts, wine critics, and marketers attach dollars to bottles. But the Toad Hollow story was never about pleasing these dollar-dishing chumps who cackle about how much they spent on a bottle. Everyone knows there are good wines, great wines, and occasionally wines we use to goose the drain with. High prices for some are justified; rarity, vintage, and best in class will always command respectable prices.

It is one of the few ironies that follow the business of wine — that one must pay mightily to consume what is best. For many, drinkability has become a price-influenced matter. The idea that taste is greater if you pay more is a wine fallacy based on consumer mindset. The critic who spoke of "respect" regarding Toad Hollow wine was referring to cost and saleability, not taste. It was a reference to consumer perception and not the wine's quality— not a note about its personality or the influence of its terroir, but a price tag glued to the bottle.

The more than 100,000 cases Toad Hollow has distributed across the country is a tribute to Todd and Frankie's story; a success story based upon a model of affordability, fun, and quality without price tag worries. It is a legacy that continues today and is shared at the Toad Hollow tasting room, centered in the town of Healdsburg.

Outside of Healdsburg, west of Highway 101, is the Dry Creek Valley, a stretch of road as rural today as it was nearly 40 years ago. While the wine industry continued to boom in the '70s, I was spending free time chasing after the legendary "Volkswagen" (a wild boar in the Frei Brothers Vineyard), or fishing Passalacqua's private lakes near Healdsburg airport. Those days I was more concerned with oversized pork chops or trophy bass than what wine would be best for the pairing.

I met up with Paul Bernie at his Dry Creek Valley ranch and vineyard. Paul's a life-long Dry Creek farmer and self-described sharecropper — a vineyard manager who worked the region for decades. Years of farming takes a toll on the land over time and when a vineyard is depleted of nutrients, there are few options that can help its demise.

This is when human influence can best complement a vineyard: through soil enhancement. Many growers "fertigate" to help bring needed nutrients to the vineyards. Fertigation is micro-irrigation with water that is fertilized. It's a common practice among many growers.

When a family takes over the parents' farm, or a farmer's been advised the vineyard land is spent, there is need for soil nourishment.

Some owners prefer an organic approach to replenishing depleted nutrients rather than the chemically-infused method of fertigation. One sought after method is organic mulching.

This is when they call *the fixer*: Paul Bernie.

So off we went on a hillside roller coaster ride to visit a block of vineyards in `Paul's clunky white Toyota truck. He offered to show me some results near Raymond Burr Vineyards, which was established in the mid-1970s.

The live oak tree branches raking at the window had me bobbing and ducking as we circled the sloping vineyards, now rejuvenated by his mulching. Though again, water is an issue here. The table was low. The owner could drill deeper but would tap into the chemical element boron, which in heavy concentrations is poison to any vineyard. When I asked Paul how the drought has affected the farmers in the region he simply stated without whining, "Wet or dry, we deal with it."

And just like the farmers deal with the water, they deal with depleting nutrients in the soils. On this particular block, Paul told me the family had chosen not to take out the existing vineyard. The soil was exhausted but the vines were sentimental; this agricultural keepsake was once planted and cared for by their grandparents. There are many occasions when another generation cares little for the passing down of things; it's considered "old stuff." There is a lack of interest for some in the passing down of fields and those vineyards adorning them. They sell or often rip them out and plant new vines. Not here. This was a family whose thoughts of their mother and father mattered. Their parents had once cared for and worked these vines with the same attention they had when raising their children. They felt a kindred spirit to the land. So with the vines ailing, they consulted with the most knowledgeable professionals in the field, with hopes of saving them.

And this is Paul Bernie's sweet spot.

Paul's method of organic mulching isn't something new or trendy. He's been taking pomace from local vineyards and oyster grindings from nearby coastal farms for years. When he took me to visit his vacant lot between Dry Creek Vineyards, there was a whale-sized pile of crushed oyster composite floundering between vineyards. A "mixologist" himself, Paul was waiting for the pomace to start coming in so that he could prepare his calcium fortified concoction. "I use 500 tons a year," he said. Then he pointed to the beast: a tractor with a contraption that looks fitting of a Star Wars movie prop: a mulching auger jutting abruptly at a 90-degree angle from the tractor's rear.

After a moment of contemplating this bone crusher, we headed back to Paul's truck. He showed me where his wife plants vegetables each season and sells them to the local markets. As he pointed to the area I noticed he was wearing sandals. I'd heard from locals about this guy whose been working in vineyards his entire life — pruning, advising, consulting for RRV grape growers — in sandals. I was intrigued.

We've all heard reference to working individuals as "boots on the ground." Some grape growers today are referring to workers in the vineyards as "boots in the field." So naturally I was intrigued by this bare foot approach. This was "sandals in the vineyards."

The notion reflected a more personal, meaningful approach to working the vines. Certainly the loose soil, dust, and moisture influenced a greater touch with the earth. The feel of insect life upon his sun-browned feet and vegetation between his knuckled toes meant a closer association with the land. I wondered if there was a systemic connection, some biodynamic association or organic meaning behind Paul's barefoot approach. Maybe he was onto something here. So when I asked Paul — with great anticipation about this connection with the earth — what was the reason for not wearing boots all these years, he replied quite frankly, "My feet get hot."

Okay. Well, whoever said "never let the truth get in the way of a good story" hadn't shared it with Paul Bernie. I left my visit here knowing that with Dry Creek Valley farmers, there was little embellishing what happens in the field.

Another sub-appellation of the North Coast AVA, just south east of the Russian River Valley, is the Sonoma Valley. It was here at St. Anne's Crossing, formerly the site of St. Francis Winery, where I learned the craft of winemaking alongside their first winemaker Bob Robertson. This was during their boutique days, long before the winery moved down to bigger digs. It was a time when an emerging variety called Merlot was coming into its own. I was introduced to Andre Tchelistcheff and Bruno Benzinger. My sister met her future husband while working here, and it was Joe Martin, the founder of St. Francis, who graciously provided the winery for their wedding reception. I now have three amazing nephews due to this communion of lives whose ultimate journey began here.

The Sonoma Valley itself is rich with history. Early missions like San Francisco Solano planted vineyards here in the 1820s. One of California's first official wineries, Buena Vista, was established here in 1857 and remains an icon today. Further west along Highway 12 is Chateau St. Jean. Constructed in the 1920s by a wealthy iron and lumber Michigan couple, it became a founded winery in 1973. Today, Chateau St. Jean's high-scoring Cinq Cepages is crafted by acclaimed winemaker Margo Van Staaveren, with 2017 being her 37th harvest here.

From Carneros to Kenwood, the eastern slopes of the Sonoma Valley are draped beautifully in vineyards. Names like Kenwood, Gundlach Bundschu, Lasseter Family, and Gloria Ferrer Caves and Vineyards are among the many fine wineries that can be found here. Yet few speak of Sonoma Valley's history more than the Kunde Family Winery. And no man so expresses its remarkable past and view of its future than Jeff Kunde.

As Jeff and I stood on the Mountain Top overlooking land he and his family has owned in Sonoma Valley since 1904, he pointed towards the hazy distance, "Right there is San Pablo Bay. On a clear day you can see the city of Oakland." He then began identifying AVA locations like a wine country docent, his voice throttling with enthusiasm. "Petaluma Gap's there. Over here is Carneros. Right there's Moon Mountain."

If I didn't know better, I would've thought this was his first time sharing the experience. Of course, he's probably shared it with more people than Cooper — his Australian cattle dog — has chased Jack rabbits from vineyard rows. As we took a moment breathing in the view, Jeff turned and looked towards a nearby hilltop, his words now reflective. "Over there we spread our parents' ashes," he said. "It was their wish." There was a sense of accomplishment to his words, as if the Kunde family was right there, still looking over things.

The Kunde family history, like Sonoma Valley itself, is drenched with color. Jeff's narration of its chromatic past enhanced the moment, perhaps because Jeff is a born storyteller. Passed on through generations, his genuine knowledge of local history is tireless. Standing there, I felt not only that I was on Mountain Top, I felt like I was on top of the world.

The Kunde tour ride would end on Mountain Top, but the Kunde land lingered higher into the unplowed hills. It was in the mid-1970s when Bob and Fred Kunde bought the neighboring Kinneybrook Ranch on a whim, adding 850 prime acres

Along the California Wine Trail

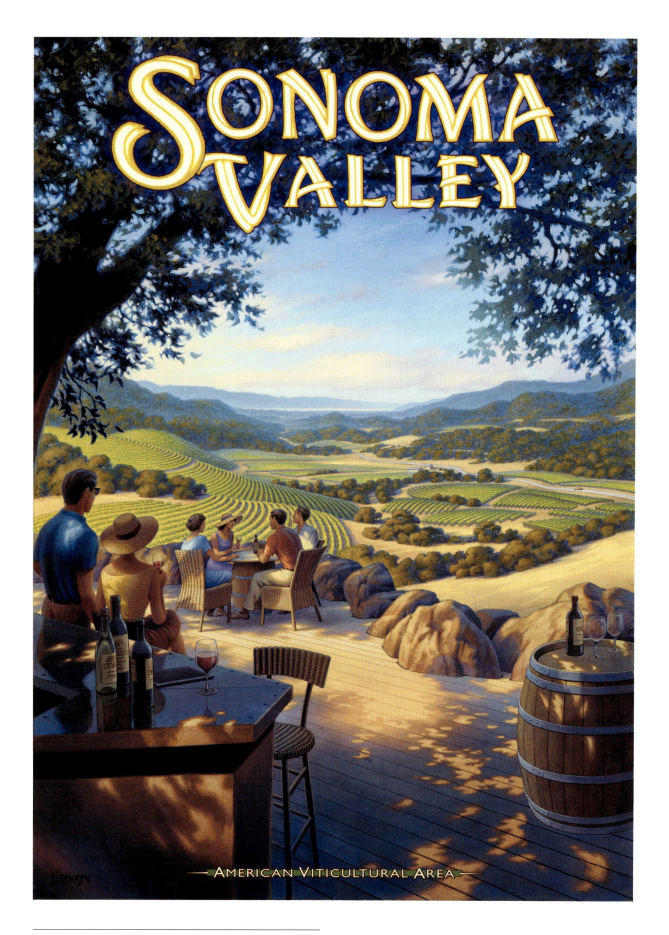

that encompass a lengthy two-mile stretch of Sonoma Valley frontage.

As we drove along the vineyard blocks above Highway 12, I could see below. The indigenous buckeye trees were in full bloom, and the California poppies were coming to life from last year's seeds. For years I have traveled this highway, passing Kunde's old barn, Wildwood Ranch, and the impressive Kinneybrook farm, where stately green magnolia trees line its entrance.

But, today, rather than traveling the highway, I found myself lost among the lush vineyard acres comfortably above the streaming line of cars. When crossing through the historic "four corners" intersection, passing by city-sized blocks of Cabernet, Chardonnay, and Sauvignon Blanc, city life seemed infinitely distant. It was vineyard wilderness drunk with lore. Its historic past is no more evident than when we stopped at the ancient Zinfandel vines at "four corners." It was here in 1879, that James Shaw and Captain John Drummond brought Cabernet Sauvignon cuttings, acquired from Chateau Margaux and Chateau Lafite Rothschild, and planted them on this spot. To put this in perspective, the planting happened the year Thomas Edison created the first lightbulb.

Though the Kunde property may not be ingeniously electrified, they have set aside a chosen spot of land that continues to light some serious celebration. Cast beneath a rising full moon, this spot might best be described as a glowing venue with a vista. Here, couples tying the knot create a union overlooking the serene Valley of the Moon. Actor Seth Rogen and other notables have been married here. The wedding grounds feature a full bar area, and arched entry, and a panoramic viewing spot. As we walked the grounds, Jeff explained how today's generation doesn't spare any expense when it comes to weddings. "One wedding planner wanted to remove the entry arch." Jeff, having built this for special occasions, said, "that wasn't gonna happen." And it didn't. It turned out the clever planner spent $5,000 on flowers just to fashion the arch up.

These hired guns do know how to get things done, especially when it comes to other people's money, like heart-smitten couples without a budget.

It's not surprising the Kunde property attracts couples. Glamour, romance, and love vibes are in its Northern California air. It's sexy and intoxicating. Yet, as we continued our drive among the vineyards, we approached another knoll where there was little romance about it. It didn't even have a fashionista's touch. It was something more fitting of a sports arena with fans screaming over a Rocky Balboa theme song; beneath the live oaks was a lumbering boxing ring. That's right, that thing where individuals bloody up their noses. There was no seating. No plausible reason for its existence. It was completely random, as if it fell right out of the sky and plopped down on the property. Although, it did deliver a knockout view.

The mind wanders terribly when confronted by the inconceivable. And mine, at that moment, was no exception. Jeff lost me here—this between Mountain Top and the tasting room spit bucket. I drifted. Could this ring be a place where those elite wine snobs step into corners to cordially sip and taste before delivering hideous wine-speaking punches? Probably not, but I do know many individuals who would pay to see it.

These hired guns do know how to get things done, especially when it comes to other people's money, like heart-smitten couples without a budget.

It's not surprising the Kunde property attracts couples. Glamour, romance, and love vibes are in its Northern California air. It's sexy and intoxicating. Yet, as we continued our drive among the vineyards, we approached another knoll where there was little romance about it. It didn't even have a fashionista's touch. It was something more fitting of a sports arena with fans screaming over a Rocky Balboa theme song; beneath the live oaks was a lumbering boxing ring. That's right, that thing where individuals bloody up their noses. There was no seating. No plausible reason for its existence. It was completely random, as if it fell right out of the sky and plopped down on the property. Though, it did deliver a knockout view.

Thankfully, before drifting further, Jeff explains the reason for the boxing ring's round-about presence. It turns out, to a lesser degree, it did have to do with punching out noses over wine matters. This four-cornered, roped and matted ring was constructed by Hollywood for a '70s-era period flick: Bottle Shock, where heartthrob actor Chris Pine gets his ass handed to him in the ring by his father, played by Bill Pullman.

Although Jeff showed as much enthusiasm for the Hollywood thing as the Mountain Top cows did with our presence, he did set my floundering mind at ease. It didn't matter that the movie was based on Hollywood's version of the Judgment of Paris (and I must say, a funny one); Jeff was more interested in getting to the owl boxes down a vineyard row. When we reached them, his enthusiasm was immediately recharged, and for good reason; Jeff's a family guy. "The owl boxes were painted by local children as part of a community outreach program," he says, beaming with a father's delight.

Personal matters hold significant value to Jeff. They're present throughout the property in small details like the cute bird boxes that rise on poles above the vineyards. The children's participation, like most every Kunde idea, is another nod to community and to sustainable practices. "Sustainability is about continuous improvement," Jeff said, which the winery wholesomely embraces. Initiatives both large and small are the winery's path to continued sustainability, such as an acorn tree planting program; construction of small islands on vineyard ponds for nesting geese; solar installation; and recycling programs.

The Bonded 202 Kunde Family Winery was the first to be certified sustainable in Sonoma Valley, and Jeff's involvement with community is a major factor in operating the sustainable train. He's the Board Trustee of historic Santa Rosa Junior College which is building a Kunde Hall in his honor. In 2008, the prestigious Governors GAALA Award was presented to the winery. Of course, there's no community involvement without the dogs. Just ask Cooper. The little guy tags along in his masters' shadow — straying only when a tasting room enthusiast cajoles an affectionate hello. And it is this affection for dogs that leads us to the once-a-month dog hike through the vineyards — an event that includes potential adoptees from local shelters. These unleashed participants are treated to a cage-free romp in the vineyards, with hopes someone might find them a caring home.

Included in the event are dog tastings. Yeah, you heard it right. Again, another random and quirky Kunde event. Certainly, slobbering water bowls infused with hints of chicken or strong notes of beef is literally over the top. But, it's an event that scores big and keeps the Kunde vibe alive.

Although the event might be begging for calamities, there are more challenging issues facing the Kunde business. When I asked Jeff what those were, I expected the usual reply I've heard from growers over the years— water, climate, government regulations, maybe disease— but it was none of those. According to Jeff, it is "competition; the wine industry is the most competitive industry in the world," he said pointedly. "Think about it: when you walk through the grocery store aisle, there's maybe a dozen or more selections of brands to choose from, whether it's cereal, breads, or other staples. Now, look at the wine selection. It's enormous."

Aside from challenges, there's the ever-evolving next phase Jeff has in the works. Just steps from the tasting room doors are the beginnings of a remarkable block of vineyards coming to fruition. "These are cuttings sourced from old Zinfandel vines we gathered throughout the county." He points over the winery's front pond. "We're building a bridge, over the pond with a tasting area here with these old vines. It's going to be the Ultimate Zinfandel Experience."

And for good reason; there is no California variety so akin to the Kundes' own story as Old Vine Zin. It speaks of past generations. The vines reach an age our grandfathers could not. Stubborn by character, resourceful by nature, and naturally less fruiting, the ancient vines have outlived immigrant winegrowers who planted the very rootstocks, dating as far back as the mid-1800s. Unlike other varietals, such as Chardonnay, Cabernet, Pinot Noir, and a plethora of others, Zin is blessed with a life of longevity, surviving not merely to see another day, but another century.

The majority of soils enriching the Kunde vineyards consist of a volcanic band of Red Hill soil from lava flows millions of years old. The trifecta of tiered wines produced from these soils include the Estate Series, Destination Series, and Reserve Series. The grapes are sourced from their estate vines, bottled, and fittingly poured in a tasting room that is remarkably all-inclusive. Here is where many phases of the winery can be seen through its tasting room windows: the vineyards whisper their hellos through framed glass windows; the glimmering stainless steel tanks stand as sentinels near the wine cave entrance; and even the bottling line behind a glass wall silently spins out an estimated 70,000 cases of wine each year.

As we left the tasting room. Jeff opened the door for a group of visitors. "Welcome," he said, smiling graciously. "Where are you from?" The crowd sheepishly replied, "North Carolina," not knowing they're speaking to Jeff Kunde. Just as the last of the group enters, a young man formally introduces himself. Jeff shakes his hand and says, "Jeff Kunde." Hearing the name, the group turned to look back, and smiled appreciably. This is Jeff in his element. He fits seamlessly into the social groove. He enjoys coming off as being just one of the guys. In many ways he plays the role of undercover boss well. Only, he does it genuinely. Jeff clearly enjoys his role at the winery, but it's the business climate that he lives and breathes. He didn't get this far by simply being chummy. He's business savvy and wine smart.

> This region has brought well known men and women to romanticize its place in history.

Although the Kunde Family empire resonates with achievements, Jeff still refers to the winery as simply a mom-and-pop endeavor. And he says it with great pride. For some this might come off as being old fashioned, but for a guy like Jeff, whose parents' ashes are spread near the Mountain Top, it's just the way he rolls.

Established in 1981, Sonoma Valley sub-appellation — championed by the Kundes — is one of California's most iconic growing regions, producing quality wines from the vineyards thriving in its fertile soil. Throughout the Sonoma Valley there are nearly a dozen soil variations; the fruit is greatly influenced by the distinct character and personality of the land. At the region's southern end is Carneros, where the sedimentary soil is marine-based. The mountainous areas and bench lands are mostly volcanic-based soils that are well draining and produce exceptional varieties like Cabernet Sauvignon, Chardonnay, Pinot Noir and Merlot.

The allure of Sonoma Valley is in its simplicity, with its rolling hills and shouldered hillsides crawling with oak. Many days and nights I've spent with good friends, captivated by a full moon rising, watching its perfect symmetry slowly launch free of a ridge top, and its glowing presence cast shadows in the night. This region has brought well-known men and women to romanticize its place in history, none more capturing its abundant beauty than Jack London, a grape grower himself, who wrote of the Sonoma Valley:

> "The grapes on a score of rolling hills are red with autumn flame. Across Sonoma Mountain, wisps of sea fog are stealing. The afternoon sun smolders in the drowsy sky. I have everything to make me glad I am alive. I am filled with dreams and mysteries. I am all sun and air and sparkle. I am vitalized, organic."

For several years, I lived in a house that bordered Jack London State Park in the Sonoma Mountains. It was a small house whose foundation was carved into the steep mountain slope. A large deck overlooked a portion of Glen Ellen and the Sonoma Valley. It was near St. Francis Winery and it was home for several years before I moved back to Santa Rosa.

Santa Rosa is Sonoma Valley's western neighbor. Its population has exploded over the years while its roads have remained much the same. If you travel north of Santa Rosa on Highway 101 and turn on River Road you can shake the traffic, but not the beauty of the vineyards! It's everywhere until you reach Guerneville. And a dozen miles west of this bohemian river town, you'll be abruptly stopped by cliffs, with barnacled rock formations and the clapping waves of the Pacific Ocean applauding you, for you have arrived.

It is the end of the road here if you're heading west, at least until you hit Hawaii. The Pacific Ocean greets the traveler like nowhere else on the West Coast. If the fog-slicked cattle guards don't stop you, the turnouts will, and for a good cause — the views are immense. The Pacific coastline's presence not only defines some of the most visually appealing vistas in the state, but defines Fort Ross-Sea View, an AVA federally granted in 2012. Its designation was championed in great part by Linda and Lester Schwartz, owners of Fort Ross Vineyards, an estate that sits closest to the ocean's steeped and treacherous beach land.

The region's first wine came from vineyards planted by the Russians, who occupied the land here

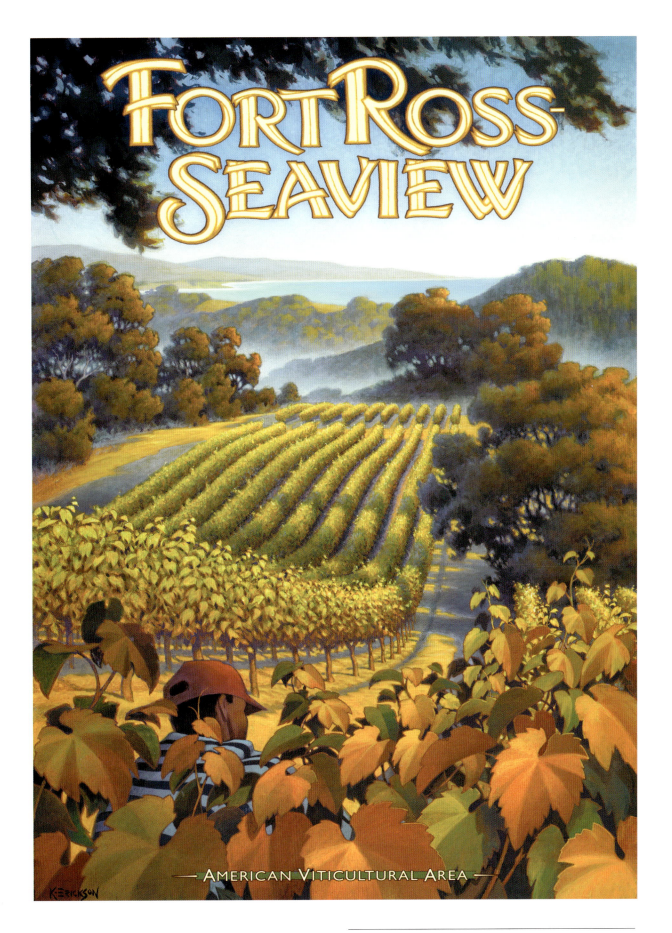

in 1812 and whose lost Palermo grape variety from Peru was said to have been the region's first varietal plantings. Since then, during the 1960s and 1970s, individuals came to the region as hippies, organic farmers, and guerrilla pot farmers. They were back-to-the-earth hippies, men and women raising communal families, nonconformists moving in among the more conventionally established cattle ranchers.

While the region moved forward, a farmer named Michael Bohan decided to plant one acre of Zinfandel. This was in 1972. There were no wine growers here because the experts told them grape growing was not possible. It couldn't be done. It's too cold. But for the nonconformist like Michael, it was more daring to dream what could be, rather than be told what couldn't. So he planted.

It was more daring to dream what could be, rather than be told what couldn't.

It was the beginning of something greater than what any grape grower could have hoped for. Bohan's endeavor was a success. This led to plantings of varieties like Pinot Noir and Chardonnay. Eventually, after decades of remote coastal challenges and naysayers, the dreams of a few have become the reality for many. And so, at bluffs above 900 feet, sunlight shines upon these vineyards reaching elevations as high as 2,000 feet.

Today, more than a dozen wineries have built foundations upon this dream. There is Nobles Winery to the north, Hirsch in the middle, and Fort Ross and Wild Hog to the south. There is something romantic about these vineyards that covet the land by the sea. There is an allure found here that goes beyond the syrupy love knots occasionally felt in one's gut when experiencing personal attraction. As I looked with Lester Schwartz from atop this steepened Fort Ross vineyard, viewing the fog-blanketed ocean below, the feeling was profound. It was overwhelming; this stunning presence, embraced by world-class wines, was pioneered by one man's vision. Today Lester is living it– a benefactor of Bohan and others who have contributed to the dream.

The region has come far since those "24 planked-dwellings" and garden settlements recorded by the first Russians. It is now a thriving wine appellation encompassing a total of 27,500 acres, with 520 presently planted to vine. The varieties most prominent in the region are Pinot Noir and Chardonnay, with a host of others including an indigenous South African grape, Pinotage– a cross between Pinot and Cinsault (a Rhône variety). The vine was created in 1925 by Stellenbosch University Viticulture professor Abraham Izak Perold. Lester and Linda brought the variety from South Africa to the region in the 1970s. At the time only 22 acres were known to exist in the entire U.S.

Lester remind me that this land was once occupied by the indigenous Pomo Indians, who thrived here for thousands of years, and referred to their home as "Top of Land." This reference to place had resonated with the Schwartzes so much that they labeled a Pinot Noir vintage in memory of the indigenous people whose lives once embraced this land– land they now call their own. The Schwartzes also pay homage to American cowboy poet and outlaw Black Bart, who was recorded as having robbed a stagecoach here. The old stagecoach trail remnants are still running through a portion of their property.

We stood at the edge of a vineyard slope, embraced by blocks of Pinot Noir and Chardonnay. Behind us, the fog was teasing the coastline at its fringes. Lester turned to me, his face shrouded by his canvas styled hat. He took in a healthy breath of air and with his charming South African accent asked, "Can you smell it?" A brief moment passed.

"The air is like champagne," he said.

This was no Brian Williams embellishing the truth. Not here. On top of this steep slope, embraced by thriving blocks of vineyards, the air is fortified by the ocean's presence. The definable notes of fog, forest, earth and sea filled my sense. We indulged in the ocean and land's presence. On this remote mountainside we were drinking from the same vast bucket of blended air the vines were breathing from. And I was feeling the buzz.

I had been awakened. My senses widened. I could see further. I felt invigorated because I was standing at the edge of something far greater than myself. I realized how sobering this place called Fort Ross-Sea View really is, and how profoundly grateful I was to share in the drinking of its Champagne air.

Upon leaving Fort Ross-Sea View I made my way through the Alexander Valley and stopped at Jordan Winery, whose off-road estate appears to be from another land — like Napa. A short distance from here, I crossed over the Russian River bridge and found myself in a sea of Alexander Valley vineyards. Over eons, this landmark river has carved its path through the heart of the valley, influencing its fertile soils and lush vegetation. As the heavy rains had passed and the flooding river calmed, spring revealed an immensely thriving Alexander Valley.

To be mindful of the Valley in spring is to embrace its surroundings and feel the heartbeat of a living, breathing presence. It comes from the clear flowing arteries of tributaries like Sausal Creek, Miller Creek, and Peterson Creek, all merging into the Russian River's wide swath of water.

Driving along Red Winery Road, I turned up a paved road and reached the prestigious Robert Young Estate Winery— more notably the Scion House tasting room where the valley and its vineyards can best be viewed from its cushy hilltop grounds.

Inside I was greeted pleasantly by the staff, but it was the outdoor infinity lawn that immediately checked me in. So impressive is the view it was easy to forget why I came here, where I came from, or for that matter, where I parked my car. Many tasting rooms attempt to capture the "money view" and many do it well, but from the Young's infinity lawn, I was viewing an elevated, intimate sense of place without fence or railing. There was no line of separation to distract my view. The initial buzz I was feeling wasn't due to the delicious estate wines, but from my senses coming to life as I looked over the vineyards producing them.

Whether standing on the well-groomed lawn, barefoot, or seated among the pergola-shaded chairs, I was indulging in the art of relaxation. The sight of flourishing vineyards quilting the valley floor and the hundred-year-old farmhouses planted squarely among them defines Alexander Valley as not only a charming place to live, but a viticultural region with some of the best California wines the state has to offer.

Alexander Valley

— American Viticultural Area —

Completed in 2018, the Scion House tasting room is fittingly built above the grand estate Peter Young first built in 1858. The swanky interior furnishings and historic photos displayed on re-purposed wood create a thoughtful pairing. Out the side windows are views of the oak-studded Mayacamas mountain range, separating Alexander Valley from its easterly Napa Valley neighbor.

It was here that I had the pleasure of meeting with Fred Young, winegrower and owner. Dressed in a Carhartt shirt, jeans, and ball cap, Fred embodies the rancher personality and wholesome farmer lifestyle. He is approachable and attentive. He is a carpenter, though doesn't consider himself one, and why not? His boat "Chardonnay" — which is docked at Lake Tahoe — was put together with the same care he puts into creating his high-scoring Estate Chardonnay.

Fred clearly enjoys his hobbies, but his passion is wine, and most importantly, his commitment to family. Sharing in this commitment is his brother Jim, CEO of the winery, and renowned vineyard manager. "Vineyard speak" is a whole new language when conversing with Jim. He's not hesitant when conveying his thoughts, and is well-educated on vineyard care, though speaks in sentences even a layman can understand — all with a congenial smile.

Both Fred and Jim are immensely successful in their wine-growing endeavors and are unplagued by vineyard envy. Being part of a large family with thirteen houses on the property and 317 acres of vines to tend, they have no time for egos. In fact, when the expression "salt of the earth," comes to mind, there is more to the phrase these men embody. They are more "grain of the earth" kind of guys.

Their legacy began over a century ago when Peter Young left New York in the mid-1800s with the notion of digging into California's rich gold deposits. Many families across the country uprooted themselves as the call became a frenzy. Peter Young may not have prospered with pick and shovel, but his land acquisition, that led to plowing and planting, turned out to be the Young family's fortune. Peter's purchase of the land in 1858 was less than a decade after Cyrus Alexander himself bought the entire Valley from the Soyotome Rancheria.

Well over a century later, the Robert Young Estate Winery was established, and the Scion House was built and named in honor of the legacy. The meaning of "scion," as defined in Webster's dictionary, is twofold: one means the descendant of a notable family, and the other is a shoot or cutting of a plant to be rooted in the ground. Fred stated his sister Susan, part owner of the winery, came up with the brand and, "It was the perfect fit."

The moments shared outdoors overlooking the Valley or walking on the grounds at the Scion House speak to the Young's generosity—an invitation to the public to share in their journey.

And what has been woven throughout their journey is a sense of pride. Their commitment to what matters stands as tall as the American flag Fred insisted on raising. The flag stands on the grounds of Scion, waving ceremoniously in much the same way Fred speaks of how far the family has come. There is unity and there is freedom. Whether it's Fred's son Robbie Jr. pursuing viticulture or his niece Rachel, now assistant winemaker, they've been allowed to follow their own path and are now the sixth generation involved at the estate.

These family values extend to the Mexican immigrants who work the estate as well. Jim speaks fondly about the field workers tending to the vineyard land. He explains, "Some workers have been working here for two or three generations. We created a plaque for one worker that honored over 50 years of service."

Though family is clearly their select endeavor, "we grow wine," is their motto. When I asked Fred if he prefers to be referred to as a farmer, a grape grower, or winegrower, he considered it for, well, just a moment. Not because he was unsure, but because in reality he is all of the above. "I'm a winegrower," he says. With 317 acres of planted grapes producing high-scoring wines, there could not be a more accurate statement. From the first Cabernet Sauvignon planted in Alexander Valley by Peter Young in 1858 to Jim Young's critical selection and development of the Robert Young Clone 17, their achievements speak of pioneering. And with Robert Young's four children — Fred, Jim, JoAnn, and Susan — driving the wine train, there's no end in sight. In 2016 the Sonoma County Harvest Fair's presentation of the prestigious "Methuselah Award" to the Youngs, honoring the family's commitment to producing delicious wines, clearly has been another stop visited at the recognition station.

The winery is not only receiving accolades for its exquisite wines, but are selling their grapes to stellar wineries like Chateau St Jean. Though, the chosen grapes they select and hand pick are produced and bottled exclusively for the Young estate label.

Along the California Wine Trail — 43

Hand picking grapes from select microblocks is a process of intense consideration and collaboration. The chosen grapes may come from micro-managed vineyard blocks like Bobs Burn Pile, the Scion block, or Clone 17. Or, they may be selected grapes chosen from among the dozen varieties planted, the most renowned include Robert Young estate Chardonnay, Cabernet Sauvignon, Sauvignon Blanc, all picked from vineyard blocks over a century after the first vines were planted.

Before leaving the estate grounds, I tried to envision the property as it once was, when acres of wheat fields and rows of prune orchards ruled the land. Much has evolved. Yet, with all that has been built, torn down, and rebuilt, there is nothing new about the Robert Young Estate Winery. To believe otherwise is to overlook the struggle and challenges that generations of wine growers like the Youngs have faced and suffered through, or more importantly, what they've accomplished.

The Young Estate's long line of achievements here in Alexander Valley is a refreshing nod to the value of age, as opposed to something new. New is the Amazon product delivered to your door, often with the promise of delivery on time. There is nothing new about the Young's endeavor, nor are there any promises that weather will be on time. Nature calls the shots. The only exception to any promise is something ingrained in the Scion legacy. When asked about the possibility of a conglomerate buying them out, Fred looked beyond the infinity lawn and without hesitation said, "It isn't about money. It's about the joy of waking up each morning, having food on the table, your health, and leaving something to family."

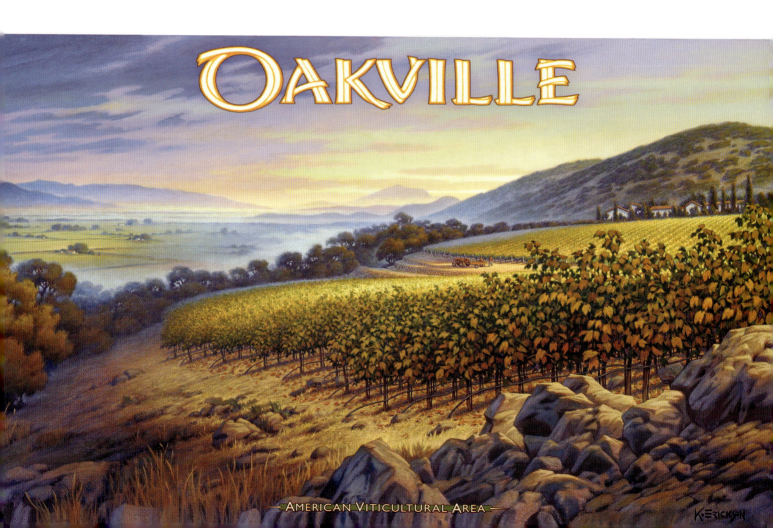

OAKVILLE

— AMERICAN VITICULTURAL AREA —

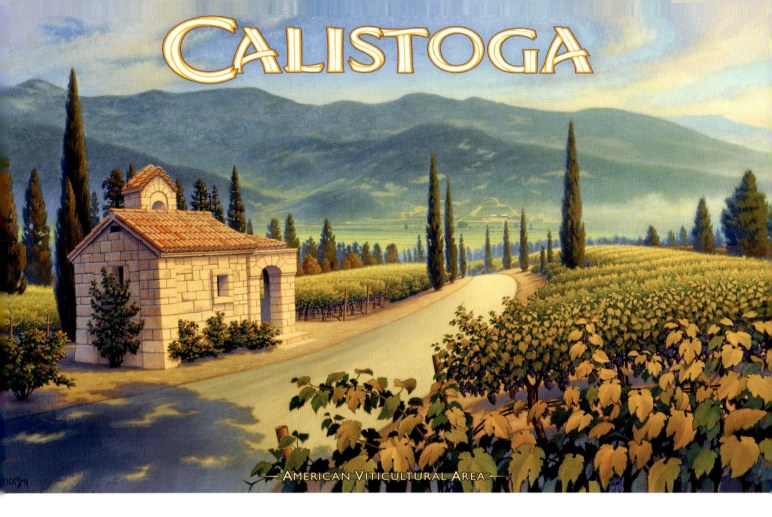

CALISTOGA
AMERICAN VITICULTURAL AREA

After an enjoyable visit with the Young family, I was back on the highway. As I traveled further along, I passed wineries like Peter Michael and Speedy Creek of Knights Valley. Turning onto Tubbs Lane in Calistoga I visited Chateau Montelena. It didn't disappoint. Its magnificence overwhelms you. I walked into the tasting room; it was swank. The crowd was sophisticated. It felt like I was at a red-carpet premiere for some tent pole movie and the buzz was not merely about the wines or its present harvest, but the stuff iconic stories are made of, like The Judgement of Paris.

This historical event was the room's centerpiece. Copies of the *Time* magazine article, written by renowned journalist George Taber, were stacked upon the shelf, free for the taking. Down the hall, a vintage of the original bottle that won the competition was displayed with ceremonial delight. Copies of accolades regarding the award, written by numerous writers, stuffed the pages of a vinyl folder. The floor director even pointed out the display video on the tasting room wall which showed clips of the movie **Bottle Shock**. And why not? This chosen 1973 Chardonnay is an abject example of the American Dream. The Smithsonian Somms are now serving the infamous bottle up to the masses for the viewing, pairing it with the likes of Armstrong's moon suit, Alexander Bell's telephone and the Declaration of Independence. I'd say that's pretty good company. I'd also say, if "some of them were my grapes" that helped to influence the wine sitting beside Lincoln's hat, I'd want some of that.

So where was the love? Where was the Bacigalupi or Russian River association? It was nowhere. It didn't exist. Not in conversation, not in framed reference on the walls, not even a whisper of the wine's true origin: grapes from another appel-

Along the California Wine Trail — 45

lation. Of course, they were Sonoma County sourced grapes, and this was Napa.

If sense of place matters and if terroir is key to the wine's journey (which I, and most every wine enthusiast will agree on), why then are the Bacigalupi grapes or its terroir not mentioned — or even referred to — as part of the story?

I asked a director at Chateau Montelena about this historical void and she reminded me that the present guests had five bottles to taste and the staff moves through them "rather quick," so discussing these issues wasn't time effective. Fair enough? I guess. Maybe. The crowd was large. There really wasn't time to get into depth regarding Bacigalupi's association– regarding the variety of significance, appellation, or the people who planted the grapes that helped bring this wine to greatness.

Yet, if the greatest story ever told about an American wine exempts the seminal elements contributing to its resounding success. And if place matters, then its beginnings must surely have relevance; not as a side kick, but with the same importance that reflects the wines most significant factor — terroir.

Certainly, Mike Grgich's craft and his master influence as a winemaker was paramount to the wine's overwhelming success. It was his wine that stumped the French nonbelievers, leaving them literally blindsided at their own masked tasting.

So, I had hoped to speak with this iconic winemaker. I heard he was in St. Helena that day, but a hard man to reach and whose health was presently under the weather. At Grgich Hills I spoke with Justin Hills, whose father Austin E. Hills is partners with Grgich. Justin and I spoke at length.

RUTHERFORD

AMERICAN VITICULTURAL AREA

Napa Valley
American Viticultural Area

When I asked about the sourcing of grapes that led to the Paris judgement, even Justin stated there were "several growers around the county claiming to have supplied the grapes." I explained to Justin that the Bacigalupis have a copy of the original weigh tag from Chateau Montelena; that in my interview with Helen, it was Grgich who had called her to announce the wonderful news; that it was her and Charles' grapes that were delivered to Chateau Montelena. I have a photo of George Taber, Mike Grgich, and Charles Bacigalupi together celebrating the news.

There is no debating the fact Chateau Montelena owned the grapes. They weighed them. They paid the grower. Their winemaker crafted the wine beautifully. They bottled the vintage and rightfully claimed the news. There's not much that wasn't theirs, other than from where a portion of the grapes had come from– picked from a block of vineyards off Westside Road in the Russian River Valley.

There is little questioning Napa Valley's stature. Its iconic presence is dizzying. With a handful of distinguished sub-appellations, it is the undisputed wine growing champion of the New World — California's vineyard centerpiece. Napa's been drawing talent and affluent personalities from around the world for years. Its acclaimed AVA draws corporate pairings to the region like that of the Culinary Institute of America in St. Helena,

Along the California Wine Trail — 47

whose past alumni have garnered chefs like Grant Achatz, Roy Choi, and Anthony Bourdain. In Yountville, the French Laundry restaurant provides world-renowned dining reflecting the very best in cuisine. The region also attracts acclaimed artists and photographers from around the globe who travel here to study its allure with lens and brush.

Does it make Napa thou art more holy? No. It means the region is further ahead of the curve. It recruited star talent. Winemakers saturated the land with vineyards. Its moment of light in this ever-changing industry shines bright, financially able to build castles and create dream lands and dazzling event centers. It also has a powerful advantage with financial backing that many regions don't. This luxury affords cutting-edge technology that leads to winemaking innovation and increased market savvy. Money matters.

And the money continues pouring into this 30-mile stretch of revered Napa Valley land.

> Its iconic presence is dizzying... It is the undisputed wine growing champion of the New World– California's vineyard centerpiece.

Stags Leap District

AMERICAN VITICULTURAL AREA

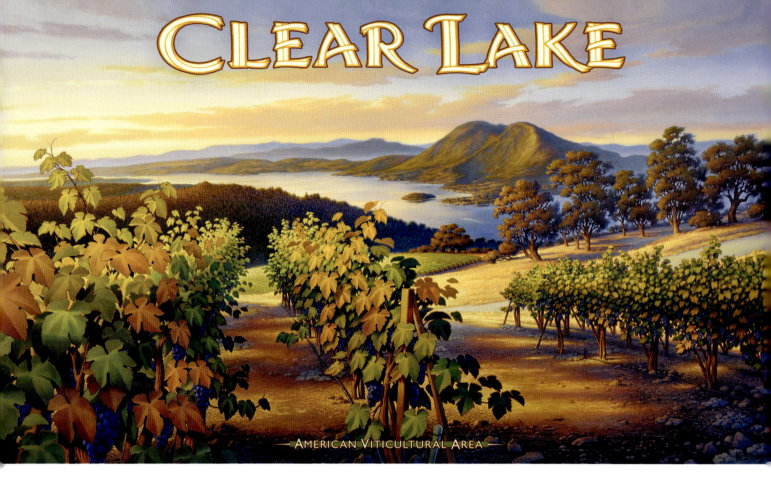

CLEAR LAKE
— American Viticultural Area —

Averaging about two miles in width, the Napa Valley bottomland and hillside vineyards grace the region with close to 45,000 planted acres. This may not be huge in numbers, but in the business of fine wine, rarely does size matter. Quality is king. It trumps yield.

Many of the grape growers reaping rewards with varieties like Cabernet Sauvignon, Pinot Noir, Zinfandel, Merlot, and Chardonnay say they farm to grow less. They boast about it. They are in search of quality, not quantity. They work in many ways to achieve this: dry farming, sourcing select fruit from neighboring vineyards, and reducing bunches per cane through canopy management. All of this serves to increase the grapes' character and flavor intensity. This is Napa Valley's splendor — taste. It has defined California as a world leader in the wine industry with a variety of celebrity names that run the course of the Silverado Trail. They are the glamour squad. The 15 iconic sub-appellations define themselves with names like Calistoga, St. Helena, Oakville, Los Carneros, Rutherford, Stags Leap District, Atlas Peak, Yountville and more. It is where talent thrives. Supporting this region of approximately 500 wineries are the many wine labs, insectaries, tech advisors, and vineyard managers. There are consultants like "the Dirt Doctor," Paul Skinner, who is so high in demand one has a hard time catching up with him. Is it any wonder? I mean, the dirt here along this Silverado Trail is so cherished (as is the region's revered Rutherford dust), I imagine they beat it out of the rugs, scoop it into jars, and truck it back to the vineyards from where it came.

And it's no wonder. When you figure in the cost of real estate here, there's some serious green invested in this dirt. The effervescent stream in a glass of Domaine Chandon aren't the only bubbles rising to the top; there's a real estate bubble happening.

In an article from March 10, 2014, The North Bay Business Journal stated the cost of a premium Napa agricultural acre in 1950 was approximately $1,000.00. Presently, the cost of an agricultural acre in Napa runs an average of $300,000– the most expensive in the country. The cost of a Silverado Napa Valley estate? In the millions. The cost of a Napa Valley winery? Priceless. There seems to be no end to the escalating property values. In fact, in the same article, the North Bay Business Journal stated that by 2050 the cost of a premium Napa agricultural acre could fetch a million.

Maybe beating the dirt out of things really isn't that farfetched.

In many ways, to travel the Silverado Trail is to experience a wine coma. Ride the wine train. Treat yourself to a walk among the mansions. Book a multi-million-dollar wedding at a wine castle. Tour the newest wine faux chateau. Soar above the Valley in the comfort of a glider or hot air balloon.

The Silverado Trail is branding itself as the Disneyland of Drinkability — a sense of place for all to share, splashed with a pairing of spectacular!

When I left the Silverado Trail it was like stepping off a cruise ship. The fun was over. The wallet was lighter. My liver was weaker and my waist-line had become measurably bigger.

It was a hangover based on pure indulgence.

Has this cruise ship feel and theme park adventure distanced the region from the intimacy it once delivered? Maybe. But who cares? Not these guys, and for legitimate reasons. Who needs intimate when grand is more, when grand is nothing short of glorious? And glory sells. While the costs continue to rise in tourist shops, tasting rooms, and on store shelves, the region is catapulting itself forward. It is suiting the economic needs of local businesses, and more importantly, its investors.

I don't think quaint is what matters, unless you ask the neighbors who live here. Growth has been an ongoing concern for residents along the Silverado Trail for years. The local newspaper, the Napa Valley Register, reports in a 2015 article, "Napa County is convening a summit to gauge if winery growth has become a runaway train." Apparently, the locals aren't feeling the agricultural love. A few could give a bung about the bottle. They're looking to put a moratorium on the local wine industry's monumental growth to better their own quality of life. Who can blame them? The traffic's gotten so bad I wouldn't be surprised if they started vetting tourists arriving at the County's border.

Yet, with all of its European flare, wonderment, and vacationland appeal, Napa remains one of the most celebrated and affluent wine regions in the world. It is so beloved by the jetsetting crowd one has to wonder, why would anyone want to fly to Europe when France exits here? Will this appellation one day be topped in prominence by another? There are emerging regions circling above the straw-hued summer hills and fettered green valley garnishing the Napa Valley — regions that may one day draw the same wealth, the same affluence and stature as the glamour squad does along the Silverado Trail.

But for today, it's impossible not to enjoy the ride.

North of Napa Valley is the dry, elevated hills of Lake County. For decades, Lake County has sourced grapes to those talented and financially-backed wineries south of them. Today, it is a wine region garnering accolades of its own. What was once a pass-through region has become a new destination.

Unlike years past, most of Lake County's grapes no longer have to travel. They can, and are, being lovingly crushed at home. For those grapes, it means no more bumpy rides over mountain borders; no more bunches squeezed into gondolas or chubby plastic bins; no more long goodbyes or rides on smoking ten wheelers through strange valleys. Sure, it could be worse. Grapes in the hands of Napa and Sonoma County winemakers with notable talent isn't a bad way to go. Just ask Grgich.

There are still wineries in Napa sourcing Lake County grapes, like Fore Family Vineyards, whose Cobb Mountain Vineyard wines are produced and bottled in Napa. For those growers whose grapes remain at home, whose wineries have become financially independent and don't rely on sources outside the county, it means local growth. By no longer having to cross the county to be branded by another vintner means growers can create a brand of their own. The region's independence has become a celebration for not only the growers, but residents who have begun to raise their glasses and cheer this much needed economic influx.

And the party's just started.

Resident grape growers throughout the county are planting with abundance. There are seven sub-appellations included in the county: Red Hills, Kelsey Bench, Big Valley District, Clear Lake, Benmore, Guenoc Valley, and High Valley, with varying soils of Franciscan and volcanic origin. Where elevations reach 3,000 feet, the grapes are protected from intense heat due to direct sunlight by a paternal force: the skin. The constant heat forces the grapes' skins to thicken, ultimately protecting the seeds, which are the procreating parts.

Local families who have been growing grapes high and low in Lake County for decades are saying, "I told you so," and new vintners are touting progress and taking the region by storm — literally.

From the Red Hills appellation to the Kelseyville Bench, the volcanic association is shaking this region. Here you'll find labels like Eruption, Lava Flow, Obsidian Ridge, and Dynamite, just to name a few. Although walnut groves, pear orchards, and pot farms are among the crops grown in the region, it is the vineyard that has taken hold of this unique land where obsidian chunks are as common as volcanic outcroppings.

In 1960, Lake County recorded about 100 acres of planted grapes; today there are close to 10,000. Wineries are rising out of the volcanic ash soils to establish themselves. Tasting rooms are bringing fresh life and unprecedented cheer.

It means more family at the table. It becomes a community connection, not only for the grape growers, but for the restaurants, cafes, and tourist shops. It is the revival of a waning economic region. Speak to most locals and they'll say, "thank goodness for the grapes."

Here in Lake County there is a colossal force of nature that rears its ugly head when the season of green hills turns into the season of brown. It is one that threatens vineyards — the region's wild land fires. Although vineyards may serve as a firebreak, a land fire storm is relentless. Its fury can devastate everything in its path; smothering vineyards, scorching vines at its fringes, and tainting grapes. Due to the seasonal association with wildfires in the area, grape growers have legitimate concerns. Harvest is likely happening.

When grape skins absorb undesirable levels of the lingering smoke, damage can occur. It can leave the grape's finished flavor crawling with notes that have been horrifically described as a "wet ashtray." Smoke not only makes the wines unpalatable, it can ruin an entire vintage. Fortunately, during the usual peak of fire season, the varieties have undergone veraison and are less susceptible to "smoke taint."

On a hot afternoon in Kelseyville at the Studebaker Cafe, I sat with Debra Sommerfield, president of the Lake County Winegrape Commission. The air was hazy outside. Two terrifying fires — the Jerusalem and Rocky Fires — had recently spared Lake County vineyards. Debra indicated the PPM (parts per million) of smoke was not an issue. She was assured by local growers that the grapes were in fine shape. Easterly winds had carried the smoke swiftly away– the winds that caused these same fires to fuel had been a Godsend.

Then, weeks later, it hit.

There was no containing what would become one of the region's worst wild land fire in history, The Valley Fire. More than 35 wineries and close to 10,000 acres of vineyards were threatened throughout the region. Langtry Estate Vineyards was spared, but the fire consumed an outbuilding, stopping at the borders of its 23,000-acre estate. Shed Horn Cellars was leveled. Growers couldn't gain access to their vineyards due to road closures. The firestorm traveled 40,000 acres overnight. Its monstrous and random activity was so unchartered, Cal Fire models couldn't predict them. It was the firestorm from hell burning vineyards, choking vines, and tainting grapes.

The Valley fire took approximately 1,000 homes and consumed several towns in less than a day. It leveled portions of Middletown, burned through Cobb along the 175, and threatened Kelseyville. The historic Hoberg House I had passed weeks earlier on my travels to the region was burned to the ground. What seasoned firefighters had deemed long overdue had suddenly happened, turning wine country landscape a dozen shades of grey.

Beyond the flames, north of Lake County, is another viticulture area establishing itself as a significant player in the wine industry — Mendocino County. This mountainous region, covers over 2,000,000 acres and includes 10 American Viticultural Areas, with more under consideration.

Scattered in small lots are 18,000 acres of planted vineyards, 20 percent of which are certified organic, the largest organic percentage in the state. This includes a large number of biodynamic vineyards, nearly 700 acres. The region is so lovingly associated with organic that those outside the county who chirp organic, look organic, or grow organic are referred to as being, "So Mendocino." The county brands itself as "America's greenest wine region."

It's also one of America's greenest outdoor pot growing regions, including itself in the notorious "Emerald Triangle." While the grape harvest is underway, marijuana bud trimmers came from across the country to Mendocino cropping the

Along the California Wine Trail — 53

MENDOCINO

— AMERICAN VITICULTURAL AREA —

sticky buds. They had for decades. Today, legalization has put an end to the illegal, but well-paying, bud-shapers job.

Grape growing in Mendocino is dependent on smaller blocks of vineyards due to the rugged forest landscape. A typical vineyard averages 10 to 14 acres. The region's best-known varieties are Pinot Noir, followed by Cabernet Sauvignon, Zinfandel, and Merlot. And as the region grows, so does the varieties of grapes planted, like Alsatian Gewurtztraminer, Pinot Gris, and Riesling.

Among the ten AVA's in Mendocino County, Anderson Valley is the most acclaimed, with its Pinot Noir rocking the region's string of hits; Goldeneye Winery in Philo provided their Pinot Noir for the 2012 Obama inauguration.

Whether you score a tasting trip among oak groves, find yourself perched on mountain tops or lounging in a tasting room shaded by magnificent coastal Redwoods, each of Mendocino's sub-appellations reflects a foggy, cool terroir uniquely its own.

Today, hundreds of wineries throughout the entire North Coast offer exquisite tasting rooms splashed with style and body. They are as diverse as the region's own varieties. Whether intimate, rustic, modern, or funky in ambience, they offer the wine enthusiast a tasting experience found nowhere else in the country. They provide space to relax, both indoors and out. From the elevated foothills of Lake County to Sonoma County, and Napa Valley, along the Fort Ross-Seaview AVA and north to Mendocino, the tasting rooms offer spectacular vistas and iconic settings, contributing to the immense success the North Coast wine industry experiences today.

For wine enthusiasts, the tasting room defines a spirit of relaxation. Its communion is not simply with the indigenous environment. They are buzzkill-free zones. They take us away from the politics and social media drama, from the "Debbie Downer" stories that suck the fresh air from us, and the ugly news stories that rock our world.

In the tasting room, they pour charm and comfort. They are ambassadors of good times, hosting and sharing in California's North Coast abundance. Like the best hotels, the wineries have hospitality directors to ensure a visit is one to be remembered. Few industries deliver this level of care and unequivocal attention. To taste the estate wines is to share the same sunshine, fog, and rainy days the local vineyards embrace. It is a place where one becomes more than members of a "club" or mere visitors within its walls; it's a place where wine enthusiasts discover moments of local distinction. Whether it's the terroir or the winemaker's craft, the character found in the glass is distinctly unique.

Personally, there have been no tasting room visits more profound than my morning visit with Helen Bacigalupi. The room was quaint and surrounded by oaks and subtle hills. The estate vineyards nearby had been cared for by Charles and Helen Bacigalupi for decades. We talked at length about the history of the winery and the region, and when our conversation led to its challenges

regarding the drought, Helen scoffed. Like most who've seen droughts come and go, Helen is a farmer in touch with place. The sunny mornings and the ebbing fog lying low over the valley have defined home for her for nearly a century. She pointed to a slideshow of historic photos on the tasting room flat screen— a black and white image of the Russian River flooding the lower vineyard faded in.

Helen turned to me and nodded with optimistic certainty. "It will come again" she said, referring to the water-laden images on the screen. Helen's words resonated with the voice of age and with wisdom that comes from life experiences. She then smiled reverently as an image of her late husband Charles appeared. Together they had succeeded in fulfilling a dream, which has been passed onto their grandchildren Nicole and Katherine.

As the slide show images continued, a seemingly unremarkable block of vineyards appeared. It was the vineyard that produced the grapes bought by Chateau Montelena that contributed to the iconic wine Grigich so masterfully crafted. In my North Coast travels, little was mentioned of their Russian River Valley vineyard. The French judges' scores have been touted. Chateau Montelena's branding of the event has been marketed. Grigich has been inducted into the Wine Hall of Fame. Hollywood promoted the movie. **Time** magazine writer George Taber gained notoriety breaking the story. What has been spoken of the wine's beginning has, at best, been only grape whisperings. Yet we know a percentage of the grapes that contributed to this seminal event were sourced from the Russian River Valley appellation; a place where barns are more prominent than castles and a seemingly unremarkable vineyard ultimately helped bring something spectacular to a Napa

County bottle. A vintage 1973 Chardonnay so esteemed it sits in the Smithsonian today, crowning itself as the wine industry's most celebrated moment in the New World.

3
THE CENTRAL VALLEY

It was an early afternoon in June when I had the pleasure of speaking with Winemaker Michael Blaylock in his office at Quady Winery. When I asked Michael what his thoughts were regarding a sense of place here in the Central Valley, he looked at me for a moment, and then with his infectious laugh, turned the question on me.

"And what sense do YOU make of this place?" he asked.

I wasn't ready for that one. This guy's good. He's also approachable and one of those guys you feel like you've known all your life. But, he had me.

We both glanced out the window towards the neighboring vineyards, and as if something else had caught our attention, we continued discussing the region's past history and present challenges. The question of "sense of place" was for the moment unanswered. And maybe for good reason.

You see, the Central Valley grape-growing region itself is in a quandary. By that, I mean it's neither an AVA nor a single region, but actually two distinct regions. Two separate valleys come together as one, whose main bodies and associated rivers meet and flow into the Sacramento-San Joaquin River Delta.

With two of the most prolific-producing appellations in the state located here — the Madera AVA and the Lodi AVA — it is considered the heartland of America's colossal wine growing region. Fanning the rural landscape across more than a dozen counties are 17 American Viticultural Areas. The land speaks of the region's suitability and of the grape growers' immense success with varieties including Chardonnay, French Colombard, Chenin Blanc, Zinfandel, Cabernet Sauvignon, and Muscat.

If you think that was a mouthful, consider this: eons ago, this entire area was an expansive sea with massive fault up-liftings and steeped valleys that evolved into fertile marshlands where prehistoric life foraged and strolled. Over millions of years, the decaying Pacific and Eastern mountains filled the valley with deposits, lifting the region to its present elevation, an average 200 feet.

Along with wine grapes, over 230 different crops thrive in this fertile soil – more than any other region in the nation. The Great Central Valley is

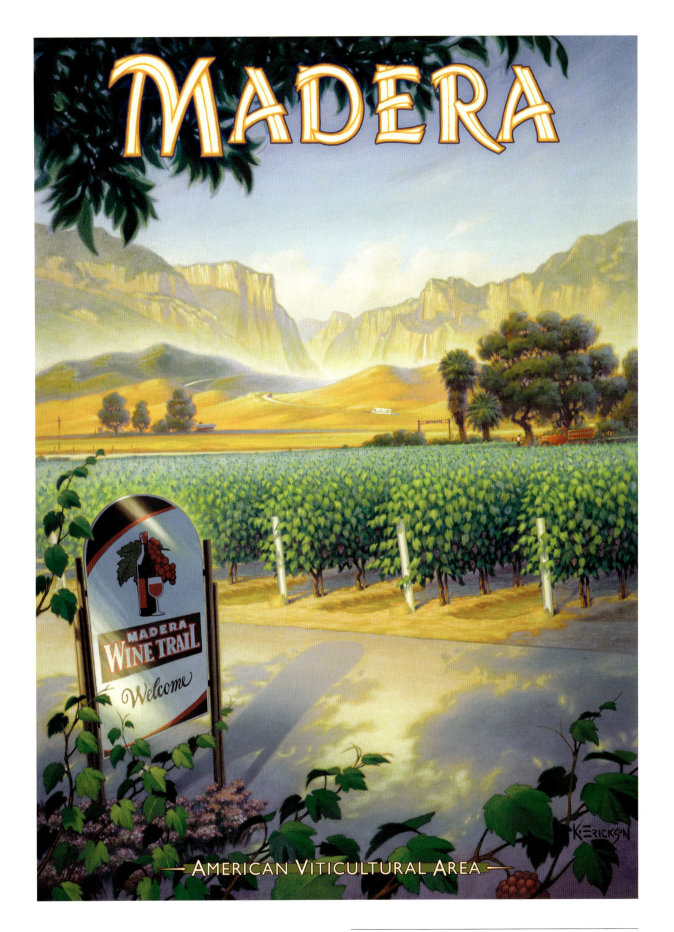

the undisputed center of California agriculture; it is the hand that feeds the world.

Farmers have been harvesting grapes here for decades. In the mid-1800s, Jonathan Holt Dodge built a large estate in Stockton, where the Holt house still stands today as Sorelle Winery. In the 1860s, the Hoffman family homesteaded 400 acres in Lodi that is now home to Heritage Oak Winery. In 1906, Franzia bought Ripon Ranch and after Prohibition came Gallo, both growing to become juggernauts within the wine industry. Among these pioneers were also hardworking laborers, water architects, and engineers whose past efforts brought sustenance to the region.

But my thoughts of traveling to this "great" Valley so drenched with history were filled with skepticism — mostly due to Highway 99. It was always that long, boring drive to see family and the long return back home. I just wasn't feeling it. It's not like it's a hard drive from Southern California. Just follow the attractive, highly poisonous, white and pink oleanders north past the dairy farms; past pot-bellied silos; past tractor stores; past pallet bone yards, industrial junkyards, and stores advertising themselves, Texas-style, like Boot Barn. It wasn't until I took the exit and left the 90-mile-per-hour crowd to fight among themselves did things quiet. Life slowed. Here, the crunch of commercial and industry leftovers faded, while the land softened and came to life with flat stretches of cultivated bliss. This is vineyard country: city-sized blocks of vineyards encompassing thousands of acres, defined by sweet varieties like Orange and Black Muscat; Port varieties like Tinta Madeira; and those table grapes we so love to snack on, Thompson seedless.

The open land welcomes you with smiling vineyards at every turn. It's approachable. Its presence shouts its grand hellos. This makes for the beauty of leaving the beaten pavement. In a few minutes, the traffic noise disappears and one is embraced by the soothing harmony of vineyards, and lots of them — to the tune of 71 percent of wine grapes grown in the state.

In the northern part of the Central Valley, the Sacramento-Delta region, are wineries like Bogle Vineyards, Gnarly Head, and Abundance Vineyards. To the south you'll find Birdstone Winery, Fäsi Estate Winery and renowned Port makers like Quady Winery. Bordering Quady is the 800-pound gorilla in the room: Constellation Brands' Mission Bell Winery.

In the outlying regions, set among the foothills near Madera, are many smaller wineries like Westbrook Wine Farm. These wineries bunch like Muscat clusters, delightfully crowding the countryside, but none define "Port authority" like Ficklin Vineyards. The Ficklin family roots go back to the purchase of the land in 1918 by Walter Ficklin and his wife Mame. Their farming began with fruit and raisins, but it wasn't until 1946 that the family began their Port endeavors in the Portuguese tradition.

I had the pleasure of spending time with Peter Ficklin and his fiancée Denise at the Ficklin Winery. Here, I discovered what lies at the heart of the Ficklin vineyard: a continuum of place and people. Just as vines grow and grapes mature, so do the best wineries and those who run them. There's no substituting experience. No embellishing the significance of family history. Not with the Ficklins. Age resonates throughout the winery, where the value of family is synonymous with wine. Ficklin Ports are arguably the best in the state, if not the country.

Embodied in the winery's Old World cellar is a living, breathing life that cheerfully haunted me the moment I stepped in. Built in the mid-1940s by Walter Ficklin and his sons, Walter Jr. and David, there is no embellishing its place as a historic, living landmark. To understand this best, one needs only to look upon the adobe brick walls that are constructed from the very soil the cellar stands upon. It is the same earth that sustains its vineyards, that houses the barrels, and ages the Port.

> Embodied in the winery's Old World cellar is a living, breathing life that cheerfully haunted me the moment I stepped into it.

This iconic work resonates with California's past. Even the decade-old ivy, whose arms grip the exterior walls, begs to call the cellar its own. The ivy vines climb and cling– they find nourishment from these earthen bricks. The trunks are old growths themselves.

Inside, I felt a pulse that resonated with something more, something grand and timeless. In this dark, dank environment, where cement floors are patinated fittingly with the scent of Portuguese varieties is where human influence begins the journey of fortifying and finessing the Tinta Madeira Port. The wine is put to rest in barrels and grand oak pipes, characterized by their quarter-sawn faces, oval stature, and hammered-looking bands. If I was born a Port, this is where I'd want to spend my time chilling out: aging with my best friends and family by my side; growing old with the best of care; and eventually liberated from this womb of darkness. My life would then be celebrated and appreciated by others at another time, in another place.

The cellar stands beside the Ficklin's 35 acres of vineyards planted with Portuguese grapes. It is a living, breathing place, a natural symphony inside and out. No one so personifies the selfless nature of the winery's presence more than Peter Ficklin, whose family story is as rich in history as its tawny Port is smooth.

Ficklin Ports are members of the elite solera process (aging wine by fractional blending), living wines that come from the very first Ficklin barreling. Each bottle contains history of the daunting challenges these fortified wines faced during prohibition. Public outcry and regulations enforced by the Bureau of Prohibition were arduous. The original government lock still dangles from the door where the brandy was stored. The old hand press on wheels is still in use and the old block and tackle still hangs perfectly from the cellar wall. To hear Peter reflect on these artifacts is to understand his reverence for family. It is a reminder of history and of place, a window into the Ficklin family past sustained by the use of primitive tools. Evolving over decades is this continuum of family, whose ties to the land continue to bring warmth, joy, and laughter to the table.

Closer to the town of Madera, among the vast blocks of table grape vineyards, is Quady winery, one of the larger-producing wineries along the Madera Wine Trail. The long journey Andrew and Laurel Quady have ventured on is an endeavor defined by deliciousness. They seek an excellence in their dessert wines, aperitifs, and specialty drinks you'll find nowhere else in the region or throughout the state.

If a winery's story was defined by genre, Quady's would be a romance– a syrupy love tale of passion and seduction. It is the story of a Muscat that has come of age by venturing further into the world, to a mature place where the self-proclaimed elite may view her as provocative, flirtatious, and sweet beyond measure. But to know her is to understand that she is refined, confident, and respected by everyone who shares in her company. With names like Electra, Deviation, Essensia and Vya, she reflects a taste more glamorous and fitting of a runway model than a 750ml bottle. Of course, she brings with her a host of luxurious companions to the table: fruits, cheeses, and decadent chocolates.

Andrew and Laurel Quady have certainly brought dessert wines to life, but it is winemaker Michael Blaylock who has, on his own terms, perfected the art of sweet. This perfection is bringing in new crowds. Millennials are educating themselves with wines like the Black and Orange Muscat, while mixologists are introducing these wines via use in artful cocktails.

Fortunately, I had the chance to sit down with Michael Blaylock in his office to talk about the region and its future. But first, I had a gripe; not with the wines of course, but with the roads. More specifically, with Madera County's forefathers who were either math majors or design engineers because there's an impersonal numbers-based road system designed to make travel, well, easy?

I just wasn't feeling it. Sure, cities have their first, second, fourth and those numbered streets in every town; it's common. But Avenue 24 1/2 and 17 3/4 and other impersonal names? You're talking about fractions. This is the agricultural capital of the world. "Why?" I asked Michael. "Why make it so difficult?" He explained to me, laughing at my ignorance, "No, it's the easiest system in the world." He went on to make good sense of why the numbers are where they are on the map. It was simple. Grade school stuff. You add this, add that, and you're over here.

Listen, I hate math. Even simple additions bother me. And when I'm driving Central Valley roads embraced by darling vineyards, I'm looking to relax and view the varietal countryside — not factor in some number to get me there.

As we sat in his office, the subject turned from simple math to a Jurassic Park-sized issue, something far more complex and colossal: water. That's right, the record drought. the state is facing is wreaking havoc, and this Tyrannosaurus Rex has reared its fat neck to bully the lives of many in the Central Valley region. When I asked Michael what his thoughts were regarding this formidable monster, in walked the water champion of the Valley and fourth generation almond grower Denis Prosperi.

"Here's your man," Michael said with his infectious laugh. It was impossible timing. This was Denis Prosperi, not only the head nut guru of the valley — almond grower extraordinaire — but the most respected and vocal Central Valley water authority in the state. A farmer whose passion for growing crops is as important as his commitment to the valley and those who make a living here. His obvious frustrations with self-interest groups and political ignorance regarding water issues today reminded me immediately of that Forrest Gump euphemism, "Stupid is as stupid does."

But Denis doesn't rant. He isn't a bellyacher or subscriber to "poor me," even as water issues threaten his livelihood. What is threaded in the jeans of this hardworking farmer is grit. Like most farmers, he's a fixer — a no-nonsense, hear-me-out, common-sense man. He's a fit fireball of a farmer, jacked in Wrangler jeans, a flannel shirt, and ball cap. No doubt the smartest man in the room. Just ask Enron, whose attempts to maintain a "water bank" (turning water into a traded commodity) were foiled by Prosperi and environmental advocates.

The shutting off of water to the valley's farmers forces new wells to be drilled and existing wells to be made deeper, ultimately causing them to weep. This can lead to mold and other issues. Michael fittingly refers to these tight growing Canelli bunches as "hang grenades."

At the end of their conversation, they were in agreement — they were to pick early. And with that, Denis departed. I felt fortunate to have met this vocal California champion and common-sense farmer. Though my visit here was short, I learned a great deal about this place and the values of the people who care for it: honesty, commitment, and genuine appreciation of the land.

Upon leaving Quady, I passed the nearby Mission Bell Winery, whose roots were built upon the Italian Swiss Agricultural Colony, now owned by global beverage brand Constellation Brands. Constellation has its own bottle factory on site. If you're wondering how many gallons of water it takes to make one glass bottle versus one almond, I didn't go there. I did leave the great Mission Bell arch to visit smaller wineries that stretch easterly towards the foothills and have established themselves at higher elevations, wineries like Fäsi and its neighbor Westbrook Wine Farm, which I had the pleasure of visiting.

Unlike the goliath Mission Bell conglomerate, with its grand arched entrance and sweeping acreage, I had to ask myself when I visited Westbrook Wine Farm, is it really a winery if there is no grand entry, and if there is no sign out front stating its very existence? Is it a winery if the road that leads you in cannot accommodate buses, wedding events, or for that matter, tasting strays from Madera's signature wine trail without an invite? Maybe not. Unless the wine made in this hidden, out-of-the-way quaint cellar is built largely upon one man's vision — a man whose accomplishments are rarely achieved in a lifetime, and whose attention to every detail applies to all aspects of the growing and making of something exceptional.

The winery cellar, hammered and anchored into the granite mountainside delivers a pleasant ambiance...

Meet Ray Krauss — winemaker, rancher, and vineyard manager — and his wife Tammy, owners of Westbrook Wine Farm. They're no strangers to wine, farming, or those involved with the Central Valley industry. Ray has worked with famed winemaker André Tchelistcheff. He has also traveled and tasted with wine icon and grocer Darrell Corti, whose involvement within the industry goes back decades.

The Krauss' unassuming cellar, with its impressive "WWF" handles on the colossal door, tells a story in itself. It's in the details. The winery cellar, hammered and anchored into the granite mountainside, delivers a pleasant ambiance with its solar-powered spring water fountain and the scent of fine wines and oak barrels. The Krausses do not open their doors to the wine crowd drop-ins.

This may be a buzzkill for the day-trippers who haven't called ahead, but the Krausses do arrange and welcome scheduled tastings.

Due to its precise spot centered on the California map, Westbrook Wine Farm may be called "California Central." Its elevation is approximately 1,500 feet. In Central Valley terms, that's high — not nose-bleed high, but above the fog — with an occasional dusting of snow each year. The morning breeze rises gently through the Fait and Accompli vineyard blocks. The air sifting through the canopies also frees the trellised vines of dust and dew and lessens the threat of mold. During the evening, winds come down from the mountains and cool the vines. Ray says, "I don't check for sugar." Sugar speaks only for the grape, not the stem. "You can have 30 brix with the stems still green, and if you pick green-stemmed fruit, you'll have a green-influenced juice." He presses the grapes between his fingers looking for color in the seed as well; brown is the ideal indicator of ripeness.

Ray and Tammy are attentive to their environment and would have gone organic, but the existing chemically treated cedar posts within their property would not qualify for organic certification. They're okay with it. They farm sustainably without certification. They don't use pesticides on their 3.5 acres of estate fruit.

Ray first moved into an existing, dilapidated green shack on the property. It was not until he worked the land, planted the vineyard, and ensured the young vines were comfortably taken care of by spring fed irrigation lines did Ray seek comfort for himself. He built a house among the oaks, where he and his wife Tammy could overlook the vines.

Ray is the evolving farmer. His lifestyle reflects the independent grape grower who has taken the knowledge of wines to a higher place. Among my time spent traveling, the independent farmer is in far greater touch with his place than the public conglomerates. A corporation's push for profits drives decision-making, leaving the company more focused on ticker tape than bud break. The grape grower is not a thousand miles away from the vineyard environment, conveniently safe in his city bungalow, in a state-of-the-art, concrete-formed high-rise where doors open for him and taxis sweep him off to cocktail parties. There's certainly nothing wrong with this lifestyle. We're all fans of luxury. I know I am. The fair question is, does this personal disconnect from the vineyard make for lesser wine? Maybe not. Gobs of money can help make quality. And having qualified people in place to manage things often amounts to success. Yet, the responsible farmer who works the land and lives upon the same soil — the very earth that runs through his fingers — is in far greater touch with the vineyards. And people in touch with place matter. The rain that falls upon the vineyard falls upon them. The light that nurtures the grape warms them. Every element that influences vine growth and grape maturation — which so richly characterizes the wine — flows through the veins of the farmers.

Few public companies share the family connection that farmers like Ray and Tammy have. But the Krauss' lifestyle doesn't make them better winemakers. It doesn't make them martyrs of simplicity. It simply reflects an honest, tangible, living character embodied in the wine. And therein lies a beautiful truth, one so often overlooked; the independent farmer wouldn't change these sacrifices for any city luxury or rooftop view. He or she wouldn't trade muddy boots for Jimmy Choo shoes; trade jeans for Armani suits; or for

that matter, trade the dirt, the challenges, and connection to the land for the most citified convenient lifestyle. This dedication to vineyard life is a deeply personal investment. Unlike a public stock offering, where a ticker price is paramount to its investors, the farmer's investment is simply living by his or her crops. There is a marriage. It comes first. I have seen it in my travels. His or her life and association with wine speaks not of buttery, jammy, or toasted notes so chirped about in the best of wines, but of unspoken notes born by living beside their vineyards. That Kodak slogan, "We don't sell cameras we sell memories," applies to the essence of wine itself. At the heart of every drink there is a moment that is shared by others. And the memories we glean from the drink are immeasurable. Do you think if wines were any less flavored, wine "connoisseurs" would quit drinking them? No. Wines are far greater than the complexities and flavors attributed to them. Certainly we revel in their complexities; we fall over ourselves by their character and delicious distinctions, but the independent grape grower and winemaker define these finer complexities and more. Their communion with place and attention to the vines are the unspoken notes left behind.

Further north, about 100 miles east of San Francisco and closest to the Sacrameto-Delta is the Lodi AVA, with approximately 90,000 planted acres. Climate, soil, and geography here is far more diverse than its southern valley neighbors. The region is cooled by Delta breezes which contributes to some tasty, hearty reds — including its flagship Zinfandel.

Transforming the dynamics of the Lodi region is the Trinchero Family Estate's massive production facility. Unlike many of the smaller wineries throughout the region, its footprint is changing the landscape and is a testament to how Lodi is becoming a prominent player in the wine industry.

Extending north of Lodi are AVAs such as the Cosumnes River, Mokelumne River, and Sloughhouse. They, too, are improving leaps and bounds in their wine quality. Zinfandel, Chardonnay, Cabernet Sauvignon, and Sauvignon Blanc are bringing prominence to this northern Central Valley region, where native oak trees define the countryside. To speak of this wine region without its association to the native oak would be like visiting California and not feeling the embrace of its iconic sun. In fact, wineries are naming themselves after their valley favorites. You have wineries like Oak Farm Vineyards, Heritage Oak Winery, Housley's Century Oak, and Oak Ridge, all wanting to be associated to this California icon and branding themselves by name and label to these trees so prominent in the Sacramento-Delta region.

Not only are the oaks hugely prevalent in the region, but they represent a relevance that reflects character, balance, and strength — elements all critical to the best wine. The oak is an absolute part of California's ecosystem. What the oak takes from vineyard space, it gives back via its company. The oak enhances soil nutrients and is home to healthy insects. Predator birds, like the Cooper's hawk and red-tailed hawk — true hunters of native rodents — perch on its imposing limbs. For centuries, the oak has afforded the region's Native American tribes, like the Miwok and Yokut, sustenance, with its abundant yield of acorns. And it once provided shade for the lounging grizzly.

The oaks that have aged for centuries stand today as sentinels throughout much of the Central Valley, reflecting the farmers toil, the vineyards sustenance, and communion with earth. There are many associations with the oak; it has even found its way into the wines. The use of coopered oak barrels adds natural tannins and structure to the wines. The tree lends itself as furniture to the table, bringing us closer to friends and family where we gather to tell stories, share dreams, and create memories that last for a lifetime.

> While I struggled to stand still beneath it, the Blue Oak stood stoic–cleansing in this remarkable bath.

On a Wizard of Oz-like stormy afternoon, as I traveled through miles of vineyards to visit Heritage Oak Winery, the air was choked with dirt and the vines shivered violently, their long canes contorting and leaning eerily sideways. The only thing missing was that creepy jingle and a witchy, bike-riding lady dressed in black with a scruffy dog in her basket. It was that scary. The winds dusting the entire region were so fierce they made headlines in the local paper the following day. I half-expected to see Dorothy on the front cover clutching Toto in her arms.

It was a great windstorm and a memorable moment for me as I was greeted by an ancient blue oak that stood in the center of the Heritage Oak winery's entrance. It laughed at the prevailing windstorm. I felt calmed by its presence and as I stood beneath it, the ferocious winds seemed to be nothing more than a fluffing up of its upper branches. While I struggled to stand still beneath it, the Blue Oak stood stoic, cleansing in this remarkable bath. It virtually howled with laughter. It spoke of fortitude and sustainability. Is it any wonder the owners of Heritage Oak, Tom and Carmela Hoffman, embrace it?

Like the oak, there is a great sense of heritage here. Tom's ties to five generations goes back to when the land was homesteaded in the mid-1800s. They've come a long way. The Hoffmans began their vineyard plantings around the early 1960s with the more popular Tokay table grape. Tom said, "The more they planted the table grape, the greater the glut." This lead to less price per ton and growers were forced to plant other table grapes that had not glutted the marketplace. It was during the early 1980s that wine grapes staged a renaissance.

Today, Heritage Oak is thriving. The farming practices that the Hoffmans have in place reflect everything that is sustainable without certification. To them, responsible farming is merely common sense. It's an obvious responsibility. The environmental and economic impact they have on the land and community is paramount.

In many ways, generational farmers are tired of being told what to do. They are an independent bunch who are constantly having to deal with issues forced on them like emissions, additives, and water usage. Regulation has become an open spigot of government restrictions, drowning practical farming with bureaucratic fluff. Certainly, some regulations are justifiable and necessary, but less government intervention is better. Today there's even talk about the Sand Hill Crane needing take-off room for flight; with all the vineyard land, that may be a problem. Will a regulation now be put in place, forcing growers to widen vineyard rows for the lanky crane to better lumber and take flight?

It is the younger farmer who generally tends to be more openly acceptable and compliant. The millennial generation has been conditioned to be accommodating to government intervention and grew accustomed to regulation. They've been schooled in the facts and laws that direct them to sign here and dot there.

The old-school farmers are more of an independent lot, whose deals were born with handshakes. They've worked the land far longer than many. They've seen what works and what doesn't. They've learned from their mistakes. These farmers are in many ways like the couple who lived together for years and never married, who didn't need a certificate to prove their love for one another. These farmers have nurtured their relationships with the land for decades, and it's working. They are there for each other and those around them. It's understood. To be there for the one you've invested in is more than common sense, it's a commitment to be there for one another for a life-time. These are the responsible farmers, the grape growers. They will die beside the land before letting it go fallow or allow the community they live in to suffer — all without certification.

Tom, who practices sustainability, is well aware of his own responsibilities to the environment. He says, "We don't need a little mark on our label to show we're certified. It doesn't sell wine." He goes on to say, when a tasting crowd comes through, "they don't look for a yellow mark. They're looking for good wine." This coming from a fifth-generation farmer whose family has been cultivating grapes on their land since the mid-1960s and producing exceptional wines with longevity in mind. It's a sensible practice without the sustainable certificate to frame upon the wall.

Yet Tom's son, who once worked for Lodi's certification program, Lodi Rules, is a firm believer in Sustainability in Practice (SIP) certification and believes his father should be a member of the SIP program.

Tom says respectfully and with a smile, "He and I disagree."

And disagreements will continue to challenge the Central Valley — both the north and south — wine industry's future. It's the nature of business. This vast stretch of wine country, whose jug wines juiced the vats and filled the pocketbooks of larger wineries for decades, has found its way. The money is staying at home, leading to wineries that produce finer wines of recognition. The region has evolved and is no longer stigmatized by its former bulk wine image.

Leaving the Central Valley was to understand time does not stand still. Not for the farmer, not for the vineyard, and clearly, not for its wines. What lies ahead for the Valley will be defined by future generations whose continued planting and harvesting will not only benefit the grape growers, but the community, the state, and world hunger itself.

All it needs is water to survive. There is an incredible legacy here that speaks of past generations who have made a tangible, organic, and breathing life from a vacant, parched and fallow land. The canals, levees, and aqueducts drawing from the rivers are Central Valley's lifeblood. Take the water and you take the farm, the food, and the wine; the heartland regresses, returning to its former self, becoming dry, distant, and fallow. The architects and engineers of Central Valley's vast water system, and the farmers and laborers, are responsible for bringing life to the land here. They are the true champions feeding the nation, providing food and drink to the masses. They have cared for the valley for decades, many going back generations. They have saved it from drought, disease, pestilence, and flooding. And those who believe in the saying, "They cannot save it from fools," have not met Denis Prosperi. Agricultural life will be sustained because of people like him.

As I traveled further south along the "Oleander Highway," where earlier in my travels, that boring sameness seemed to drag on for days, the time passed as fleeting as the vineyard views passing by my window. Monotony was replaced with a sense of wonder. The thought of what I was leaving behind was no longer a quandary. The region speaks of change and abundance. I thought of the mystery of this Great Valley, the promise and the fortitude of those men and women working the land here. A vast land still intimately defined by the farmer.

It led me back to the late-morning visit I had with winemaker Michael Blaylock and the question I had asked him that he so fittingly turned on me:

"And what sense do you make of this place?" I was now certain; to make sense of the Great Central Valley is to recognize it as the heartland of California. Its climate, soils, and iconic sun are unique to all the world, where vineyards thrive and agriculture grows in abundance. It is a land defined by the joining of two great valleys, sustained by the best of two main rivers that come together as one.

Here I was in the heart of it, driving the long road home.

4
The Central Coast
American Viticultural Area

When Geoff Rusack's Cessna Caravan touched down at Santa Ynez airport, there was cause to celebrate; another Santa Catalina Island harvest had successfully been delivered. After having crossed over 22 miles of the Pacific Ocean, the grapes would now be offloaded and trucked to Rusack's Ballard Canyon winery, where under the guidance of Winemaker Steven Gerbac, the vintage would be produced and bottled in the Central Coast, the largest AVA in the state.

From Santa Barbara County to San Francisco Bay, the Central Coast AVA stretches along 250 miles of California coastline. There are approximately 100,000 acres of planted grapes throughout nine counties: Alameda, Contra Costa, Monterey, San

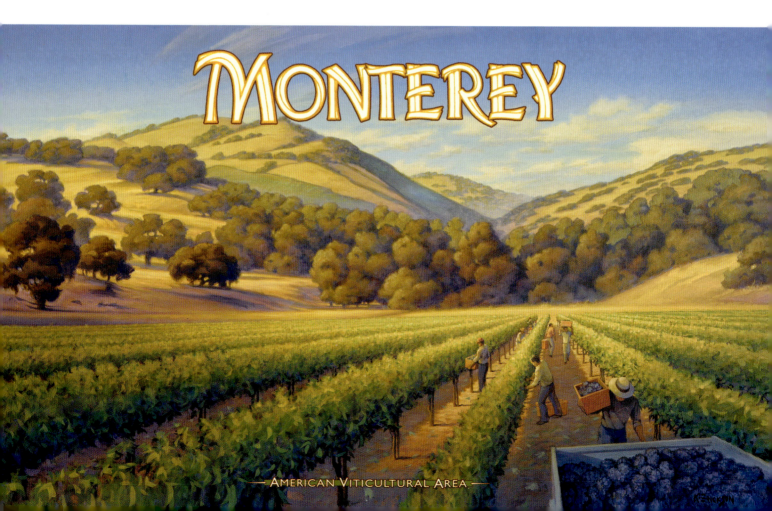

Monterey
— American Viticultural Area —

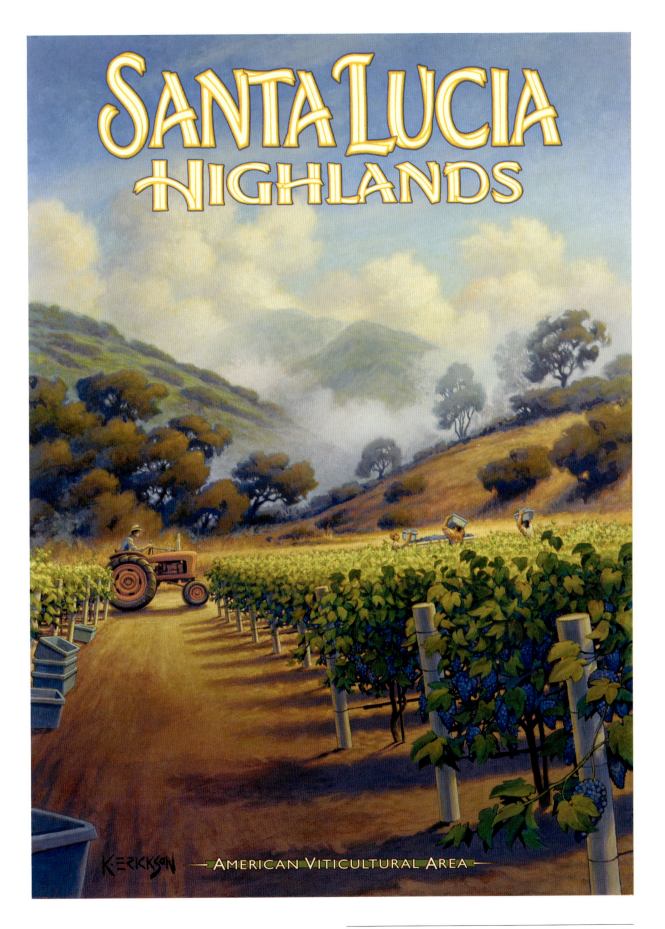

Francisco, San Mateo, San Luis Obispo, Santa Barbara, Santa Clara, and Santa Cruz. The region's maritime breezes and soil — enriched by former seabed floors — make for the perfect pairing. The terroir is influencing a diversity of varieties and producing world-class wines.

Spanning the entire Central Coast are renowned sub-appellations like Arroyo Grande Valley, Paso Robles, Santa Clara Valley, Santa Lucia Highlands, Sta. Rita Hills, Happy Canyon of Santa Barbara, and Santa Ynez Valley — a region sporting more "Santas" than a New York shopping mall at Christmas. The historical significance of the vineyards here is as rich in early California beginnings as Bordeaux's Old World roots run deep. Missions like San Borromeo de Carmelo, Santa Clara, San Luis Obispo, Santa Cruz, Santa Ynez, San Juan Batista and Monterey were producing sacramental wines by the barrelful; the vines shared the same piney breezes and maritime air as those red-cheeked friars experienced back in the day.

The establishment of these missions along the Central Coast provided more than mere places of worship; they also attended to the soldier and traveler with food and lodging. To meet service and labor needs, the church enslaved local Native Americans. They worked the mission farms, produced bricks for structures, and made saddles and soap for Spanish soldiers. They also worked the mission grounds: planting, pruning, picking, and stomping the mission grapes. Perched on elevated platforms, entire Native American families would crush the grapes as the juice drained down into cowhide bags.

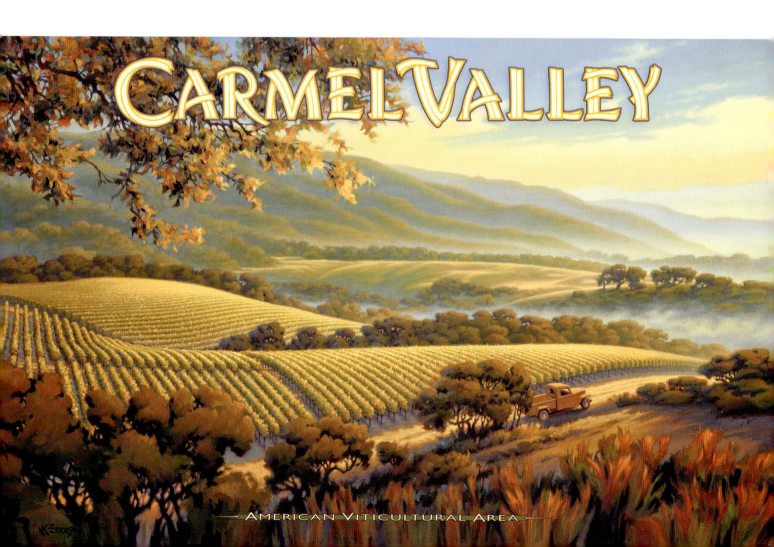

Carmel Valley
— American Viticultural Area —

The Missionaries who established vast acreage for their own unknowingly became part of a monumental wine growing region that is now embraced by wine enthusiasts from around the world. Today, nothing resonates with the wine industry's mounting success more than the Central Coast. Due to its vast, plantable acreage and fitting terroir, vineyards are thriving and the region is scoring big.

Driving toward the center of the region, passing the vast crop rows so abundant in the Salinas Valley, I could see the Santa Lucia Highlands vineyards in the distance, with the terraced bench land and hillsides rising in elevation to a height of approximately 1,200 feet. In order to taste the acclaimed wines from here, one has to drive over these mountains to Carmel Valley. It's a different vibe tasting wines from a land not surrounded by vineyards of its own. Yet, it's becoming the way to go for many wineries, some even establishing their tasting rooms in major cities.

I was best able to understand this "remote tasting room" success while visiting with former cattle rancher turned farmer John Boekenoogen. We met at his tasting room in the heart of the Carmel Valley tourist district.

John comes from a long line of cattlemen who have lived in the Santa Lucia Highlands region for generations. He never had any intentions of planting vineyards until the day a buyer showed up on his mountain property and wanted to pay "good money" for a chunk of land. Flags flew. He told this prospective buyer, "the land isn't for sale."

And why the interest?

Sure, from the mountain tops it's a great view of the Salinas Valley, where rows of broccoli and 30 shades of lettuce poke their leafy heads out of the ground. But who wants to pay big money for a panoramic view of vegetables? Not to mention the afternoon winds that blow through. If I pay for land, I want a coastal view, not a view of the "world's largest salad bowl," as John refers to it.

So John questioned the interest of these buyers and discovered that his southeast facing land, consisting of gravelly, sandy loam, was ideal country for something more than cattle: vineyards, of all things. After some negotiating, the two sides came to terms. They were allowed to plant, but not to buy. And for 10 years John sold his grapes with one stipulation: if he were to build a winery, he would be allowed to use his own grapes.

The rest was history and his family followed the winemaking call. Boekenoogen built a winery and his son Garret eventually became winemaker. His daughter Kate continues to work on the ranch and in the winery, while his second daughter Holly is involved with marketing the Boekenoogen label.

John speaks fondly of his years cattle ranching and row crop farming, though his heart is now in his winemaking. While reflecting on those days he said, "You can take a head of lettuce, hold it in your hand and say, 'Hey, check this out.' But it's a head of lettuce. Hold up a glass of wine and you've got something."

This "something" resonates with those who appreciate Santa Lucia Highlands wines. The Santa Lucia Highlands AVA was established in 1991 and stretches along 12 miles of bench land. John pulled out the AVA map and pointed to the region's distinct geography. Due to Monterey Bay's southeast air flow funneling through the mountain pass, the vineyards here are recipients of cool mornings and sunny afternoons, allowing for a lengthy growing season.

As John looked with admiration at the crowd filling this off-site tasting room, I understood that successful marketing delivers where the people are. In this case, beautiful Carmel Valley. John Boekenoogen has created a legacy here. It hasn't been easy. He makes it clear the business of wine comes with hard work and many challenges, and no one better understands the wrangling it takes to get here than this man.

Recently a land with few vineyards, the region now boasts approximately 6,000 planted acres that produce world-class wine — this due to the grit and hard work by growers like John. He's proud of his success and it shows. "Of course," he says with a telling smile, "you're only as good as your last vintage."

And while each vintage will inevitably have its own set of challenges, the regions themselves continue marching forward. Take Paso Robles, whose 700,000 total acres of land is a Central Coast behemoth. Its footprint is Saquatch-esque, with 32,000 planted acres and growing — this while consistently producing wines of acclaim.

One man renowned for his contributions to Paso Robles' success in the business of wine is Gary Eberle, considered the Godfather of the AVA. Gary was not only the first wine grower to plant Cabernet Sauvignon in the region and the first to introduce Syrah into the country, he also helped champion the designation of Paso Robles as an AVA in 1983.

I had the privilege to sit with Gary at his winery, overlooking a block of estate Cabernet Sauvignon and Mill Road Viognier. Gary talked about Eberle

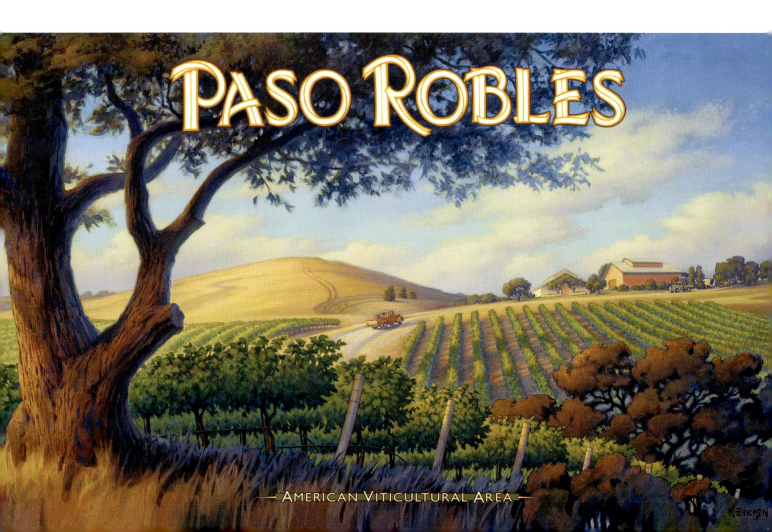

PASO ROBLES
— AMERICAN VITICULTURAL AREA —

Winery's beginnings. It was evident his passion for wine runs deep. And I'm not just referring to the 17,000 square feet of underground caves he's built, another first for the region. The wines produced by Eberle are reaching acclaim.

And while this former Penn State defensive lineman who played for Joe Paterno, speaks of the region's wine-scoring success, he doesn't sugarcoat it. There are some hard-hitting challenges, like disease and water regulations, that impact the bottom line. "Remember," Gary says, "this is an industry that prides itself on breaking even."

Of course being the "godfather" of the region and in the company of men like Dick Peterson, Robert Mondavi, and Darrell Corti, there's good reason to believe Gary, along with his wife Marcy, are achieving far greater than "breaking even." Their accomplishments clearly attest to this.

Yet beyond Eberle Winery's accolades or any number of wine scores, there is a humbleness to Gary's demeanor. There is a profound sense of comfort that resonates with Gary at his vineyard home. Maybe it's being couched on his 65-acre estate, surrounded by land he's worked and cared for since the early '70s, that allows him to feel so at rest.

Although one has to ask, when does he find time to rest? Grape growing is hard work. There's always something to tackle. And while he faces his challenges head-on, he exudes a sense of connection to the land that speaks of a kinder, more gentle man — not a linebacker knocking helmets on the turf. The more he spoke of caring for his vineyards and tending to the cork oak on the property, the more I understood what was going on. This is not merely a business to Gary. The land and the vineyards transcend personal value and reach further than his achievements.

> Here was the heart and soul of a man who wasn't leaving… Where else in all the world could things matter to so passionate a man than his vineyard home?

As he continued talking about what the land meant to him, to his family, to his employees, and to the community, he hesitated. Turning to the vineyards, he said: "I will die on this property."

The words caught me off guard. It took a moment to actually understand where he was coming from. The fact that they were even spoken during this meeting may have been awkward for some, but it wasn't for me. His words were unguarded and genuine. They resonated with what this property honestly meant to this gentleman farmer and accomplished winegrower. It was in every way the ultimate nod to the idea that place really does matter. Not just what elements are found in its dirt or what next year's change in climate might bring, but how the land is connected to the very backbone of Gary himself.

Here was the heart and soul of a man who wasn't leaving. This was home; home for his wife Marcy, their two standard poodles Sangio J and Roussanne, and home to those who worked here. They're all family, all part of a sustainable connection to "place." Where else in all the world could

things matter more to so passionate a man than his vineyard home?

The days ahead for Gary appear far more modest than most pioneers in the wine industry. "My dream of retiring is to simply pull up a chair at the tasting room entrance," he says. That's it. To greet those coming to taste his award-winning wines.

I couldn't help but be impressed by this man whose aspirations have helped him climb the mountain of success, where the air may be thin, but there's no need for oxygen. Not up here. The land is what he lives and breathes.

He has only to "pull up the chair."

Just a short poke up the road from Eberle Winery, there's an old stagecoach stop. After over one hundred years, it's now the Tobin James Cellars winery and tasting room.

There's a promise here that the American west still lives. Pushing through the front doors, you'll find everything to kick up your senses: great taste and an excellent staff graciously serving wine from behind three western-style back bars. The oldest bar was built in 1860 and came straight from Missouri.

The crowd noise carries out the front doors on almost any given day. It's not the noise of cowpokes, rough riders, or gunslingers dropping in for a snort. There will be no hold-ups other than a line of purchasing guests at the cash register. The sound is of a crowd, or posse, taking in the vibe. You get the same feeling the moment you push through the front door; that feeling one experiences of being removed from daily life.

I had this feeling while sitting with Claire and Lance Silver on their old western-style balcony above the tasting room. It was in 1996 when they partnered with Tobey James to become co-owners of the winery, which was originally established in 1993. They never looked back. Their loyal followers, which include a massive number of club members — perhaps the biggest in the world — is astonishing. "We have 30,000 club members," Claire states. Lance nods that it's true. He doesn't add puffery to these numbers but will remind you, "This is without a marketing staff."

And that's big. Claire points out that "our club members are our sales staff." This should explain a thing or two about loyalty to their wines. Apparently, club members posse up to get the word out — and it's working. At the forefront of this massive host of members are the challenges of taking care of them. To do so, they've got to have award-winning wines and hospitality that is second to none — not to mention a shipping staff with an Amazon-sized record of delivery success.

Belly up to one of their bars and you'll experience the extent of this hospitality. Personalities are found in the glass; James Gang Chardonnay, Fat Boy Zinfandel, and Blue Moon Reserve Syrah are just a few of the fine wines you'll meet here. The signature Tobin James Winery starburst icon sets the stage for a star-filled experience, and their slogan, "Paso Robles in a glass" speaks of the depth of their wines and the importance this region has in terms of making it big.

Before leaving the Paso Robles region there's a place nearby San Simeon called Hearst Castle. It is positioned, like most ocean-facing properties, to have breathtaking views. However, this estate is much larger than others. Of course, we're talking about the humble abode built by former icon publishing magnate William Randolph Hearst.

Situated within the Central Coast AVA, the castle was built in 1919 and stands today as a California Historical Landmark. I know it affords pleasure to many, but I'm not one for being herded onto buses or trailing sheep-like while being serenaded by expert tour guides. It's just not my thing. However, the opportunity to see this California magnate's over-the-top crib was too much to pass up. So as the former cattleman Boekenoogen would say, "I bit the bullet." I took the tour.

I've learned a great deal since that tour. Not only about Hearst's wealth or media empire, but about his interest in wine; it was his taste for comfort that intrigued me. It's rumored he spoke to his architect Julia Morgan about being tired of going "up there" and sleeping in tents on his property. "I'm getting a little too old for that. I would like to get something a little more comfortable," he is quoted as saying. Right? He wanted something more comfortable than a tent, so apparently he got to thinking, why not build some walls and adorn them with ancient church parts and tapestries from around the world. Maybe rather than dipping his toe into the seasonal brook, he could have an Olympic-sized swimming pool to do cannonballs in. Rather than those awkward fold-up chairs, why not some pillowy furniture from Henry VIII, or whatever. Oh, and those canvas

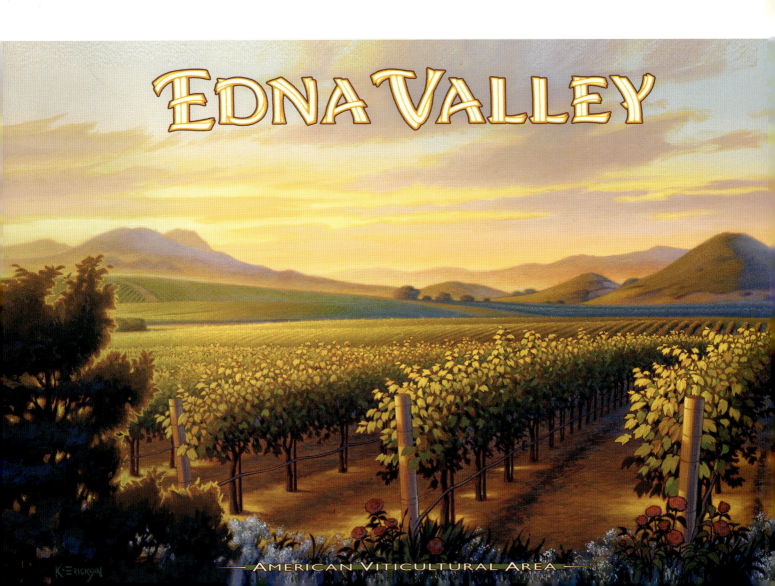

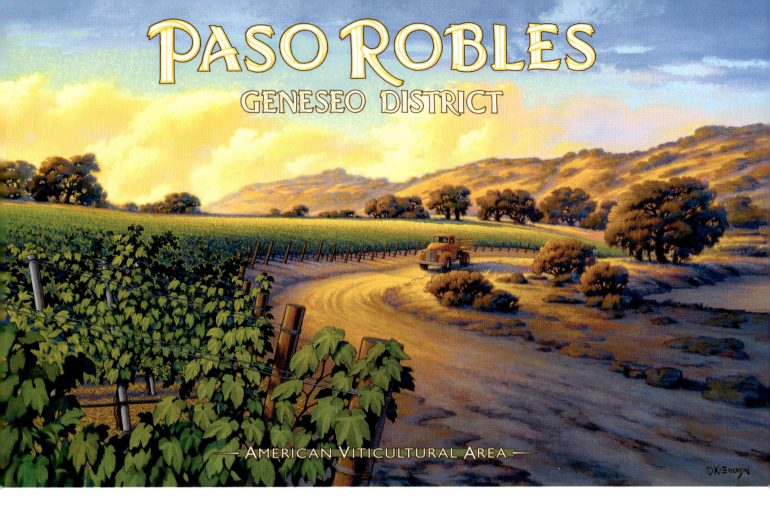

PASO ROBLES
GENESEO DISTRICT
AMERICAN VITICULTURAL AREA

army cots he slept on — forget it. Let's maybe get Napoleon's bed and marbled stone shower rooms from King Tut's tomb. Perfect!

But wait, he must have thought, let's just make it a castle. One with a fully-stocked wine cellar. This might be an appropriate substitute for a Coleman camping cooler.

It is recorded that the wine cellar was constructed and stocked with few California wines. Hearst favored French and German. In fact, he served wines at his dinner parties during Prohibition. He believed temperance should be practiced, but scoffed at Prohibition. In 1929 he wrote, "I consider the 18th amendment not only the most flagrant violation of the basic American principle of personal liberty that has ever been imposed on the American public, but the most complete failure as a temperance measure that has ever been conceived and put into impractical operation."

Hearst insisted on vaulted steel doors to guard his wine cellar during Prohibition. Few knew he had this cellar, and that the keys to it were kept in his front pocket. And what raiding Fed in his right mind would insist Hearst empty his pants pockets!

Further south, beyond Hearst's San Simeon Castle, is Edna Valley, a sub-appellation with 22,500 total acres and 3,500 acres of planted vineyards. There is a story evolving here, one beyond the prominent acres of Chardonnay and Pinot Noir varieties so esteemed by wine growers throughout this cool climate region. It's the rise of Rhône— an eminent contender in the business of Edna Valley wine. Rhône varieties are finding love among the region's rippling hills and volcanic soil. And

Arroyo Grande Valley

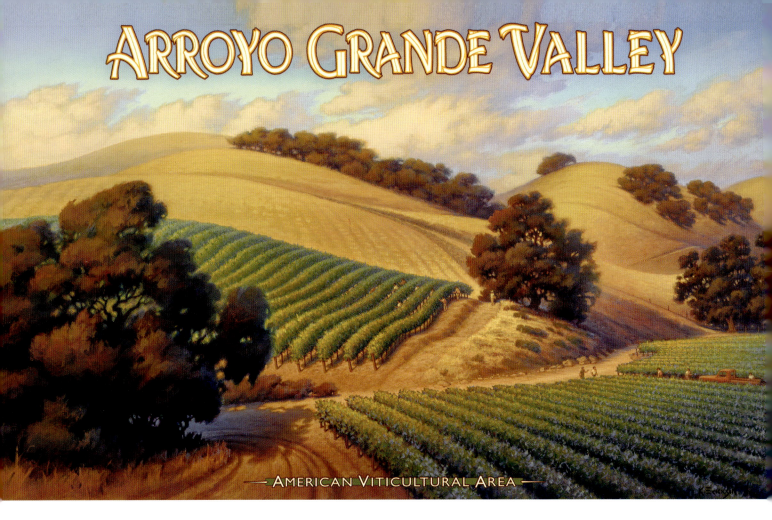

— American Viticultural Area —

there is no more a man to champion the cause than Rhône Valley pioneer himself, John Alban.

Just in from the Hospice Du Rhône event, he must have been exhausted, but he didn't show it. As we entered his office he made his position clear. "You'll find more than you need to know about me on the internet," he says without blinking. "Why are you here?"

"I want to tell your story," I replied.

"No," he said pointedly, "you want to tell your story. Mine's all on the internet."

Fair enough.

But it wasn't the internet I wanted to shake hands with. It wasn't about tapping with my fingers what a million Googlers had already done— searching and finding something significant about the man on the screen. That was the internet's story. I wanted John Alban's. I wanted to meet the Rhône Ranger himself, though I never expected to be put front and center with a fully loaded, trigger-happy scholar hungover from a weekend of hard work.

Later I came to understand that this was a man selflessly willing to give up his time, while careful not to waste it. And while John gave generously of his time, it was more than he first had to offer. His carefully guarded references to Edna Valley, his grape-growing neighbors, and people in the business were genuine and respectful, traits that flat screens just can't convey. When addressing industry issues like terroir, appellations, and varieties, he would unleash a vocabulary based on experience most textbooks couldn't keep up with. No stumbling, no contradicting. I was wondering if there was a John Alban app I should've downloaded before the interview.

I sat in the chair thinking, where in the hell did this guy come from? Seriously. I wanted a bottle of Alban wine just because he was drinking it. Syrah? Hell ya! Pour me a glass. Grenache? Viognier? I'll take a case of that. And I really had no idea if his wines lived up to their accolades. The fact is, I didn't care. I just wanted to be drinking what this guy was drinking.

What John Alban has accomplished during his affair with Rhône Valley wines is unprecedented. According to him, Alban Winery is "the first American winery and vineyard established exclusively for Rhône varieties." That's big. At his 250-acre estate there are around 60 planted acres of vineyards — all Rhône varieties. The region's average 330 foot elevation may not be high, but his consistent 90-something Parker scores are. Alban's Syrah and Viognier reflect the importance of why fit matters. He speaks of the importance of terroir first and to those varieties perfectly tailored to this place he calls home.

There is no embellishing his accomplishments. He is in every way an icon. But he will, at every turn, take you away from that and not allow you to focus on him. There is a mystery to John, something pleasantly guarded and alluring. It's not about his Aaron Eckhart looks, nor the shit detector constantly blinking in the back of his mind. It's about a similar character reminiscent in his wines, those dark and complex wines he has pioneered since 1989.

And while California continues to celebrate the grand Bordeaux and Burgundies of the region, Alban is tirelessly working to further his affair with the increased popularity of Rhône varieties, those the New World has for decades referred to as insignificant or forgotten.

When I asked John about being an icon in the business he poo-pooed the idea. "What really defines an icon?" he asked, once again directing attention away from himself. So with that, I indulged. For a wine industry icon like Alban, it's obviously more than a symbol; more than a smiley face or tiny screen symbol we tap with our fingertip; more than a Nike swoosh or McDonald's arches.

In Alban's case, an icon is one's contribution to a subject. It is an individual who breathes greater life into a matter the world is enhanced by. There is a global promise embedded in an icon. It is a promise that one's work will not be forgotten. There is no embellishing it. It takes rare perseverance and greater commitment. One does not follow without the other.

Ultimately, an icon is a standard by which the end of the tape defines the length of its measure. The greater the measure, the further the achievement, and the more pronounced is the icon. Within the wine industry there are icons like Mondavi, Gallo, and Tchelistcheff, all measured and defined by their achievements.

What has been pioneered with Rhône varieties in Edna Valley will be addressed for decades to come. It is a single legacy of a forgotten and little-cared for style left at the back of the room. Today, due to John's perseverance, Rhône Valley varieties have been allowed to come forward — no longer bullied or neglected — but now sitting at the front of the class, providing answers to questions others once deemed insignificant.

With our visit coming to an end, John escorted me out the door. We walked between the crush pad and his office, and as we stepped into the foyer, I noticed five oak doors in a circle. It was perplexing. The room was curiously shaped in the form of an enclosed wine vat — or something. And of course, I opened the door and stepped into a darkened room.

"You're not the first to walk into that closet," John said. "We call this the Frank Lloyd Wrong room." It was cause for a good laugh. As we continued our exit and entered the crush pad, there was the familiar smell of wine in production along with the sense of camaraderie among those working there. It smacked of the industry's best. Clearly, whatever John's doing outside of the foyer, he's doing it "Wright."

And you won't find that on the Internet.

What you will find near Edna Valley is a microclimate hidden in the upper Arroyo Grande Valley. It is a remote wine-growing region called Saucelito Canyon. Located south of Lake Lopez, it is seated virtually in the palm of nowhere. There is a sense of place here that takes you back in time as it defines San Luis Obispo County's wine heritage.

And no one knows it better than Bill and Nancy Greenough. In 1974, Bill began resurrecting Saucelito Canyon's three acres of lost Zinfandel vineyards that had been planted in the 1880s. It now appears these vines have discovered the fountain of youth due to one man's care and ultimate perseverance. Along with his attention comes a sense of place that matters. Imbued within the soil is "subtle gravel" and alluvial deposits — a past ocean floor releasing nutrients essential to the vine. It is a rare, orchestrated symphony of climate, soil, place and people that is producing the fine nuances present in these world-class wines.

He speaks of the importance of terroir first—to those varietals perfectly tailored to this place he calls home.

When Bill first heard of the neglected vineyard —overcome by poison ivy and tangled shrubs, and being foraged on by grazing cattle — he became intrigued. So he camped out on the property. Life at that moment changed course for Bill. His life would soon be defined by the vines he slept with, an affair that would last to this day. There was a feeling here that resonated with Bill on more levels than even he seemed to understand.

I sat with Bill on his rustic deck overlooking the ancient and flourishing vines. He spoke of the early days constructing the ranch house while bringing life back to the vineyard. The ranch house may be "off the grid" as he says, but it has been a place of family celebrations for decades. This is where he slept while rejuvenating the three acres of Zinfandel, and where he would later plant additional blocks of vineyards. Had it not been for Bill, there's a good chance these old vines would not be here today.

Yet some credit to their survival should be shared with a man named Henry Ditmas. It is beyond thinking that in the 1880s this Englishman would choose the right variety based on fit at a time when varieties were instead chosen for their popularity.

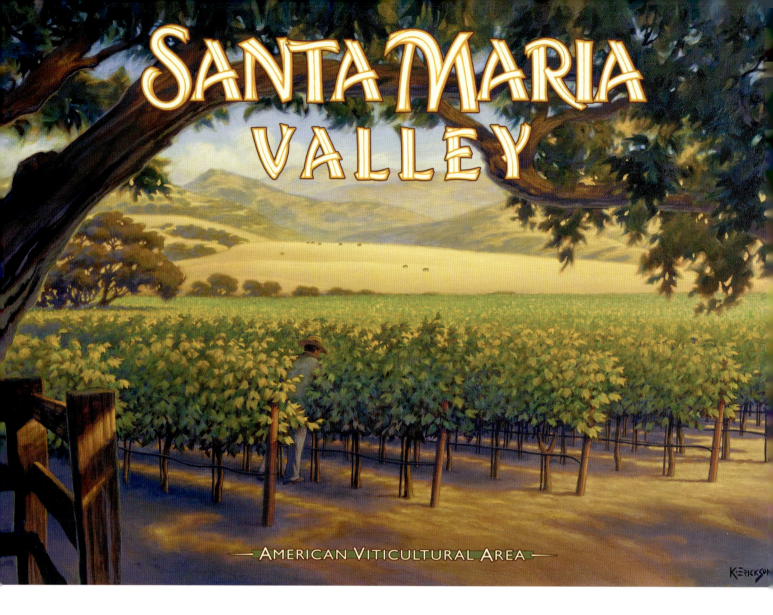

Santa Maria Valley

— American Viticultural Area —

There is a sense of nobility in these old vines. They appear to have been knighted by mother time; allowed to pass by the wrath of rot and decay, disease and pestilence, and to live another day. Here, too, the mountains appear unchanged. In the distance, Hi Mountain still captures the ancient dreamy sunsets that appear on Saucelito Canyon Vineyard's label. The fossilized oysters, clams, and silver dollars embedded along the local creek beds rest in their same places. These fossils date back millions of years. So common are the overweight crustaceans that the Greenoughs use them throughout the ranch house as attractive doorstops. They speak of a distant time when the ocean lay over the land, and the stubbing of one's toe is a blunt reminder of their petrified existence.

As I walked the canyon flats and subtle hillsides, I better understood Bill's feelings as I contemplated Saucelito Canyon's vineyard past. It is just the "eons ago" thing I shy away from. Attempting to wrap my mind around things as overwhelming or intense as how the land was formed is incomprehensible to me. It's like a math equation gone rogue. There are those professionals who know it best, but even they rely on a great amount of speculation. There is a sheer brightness to its

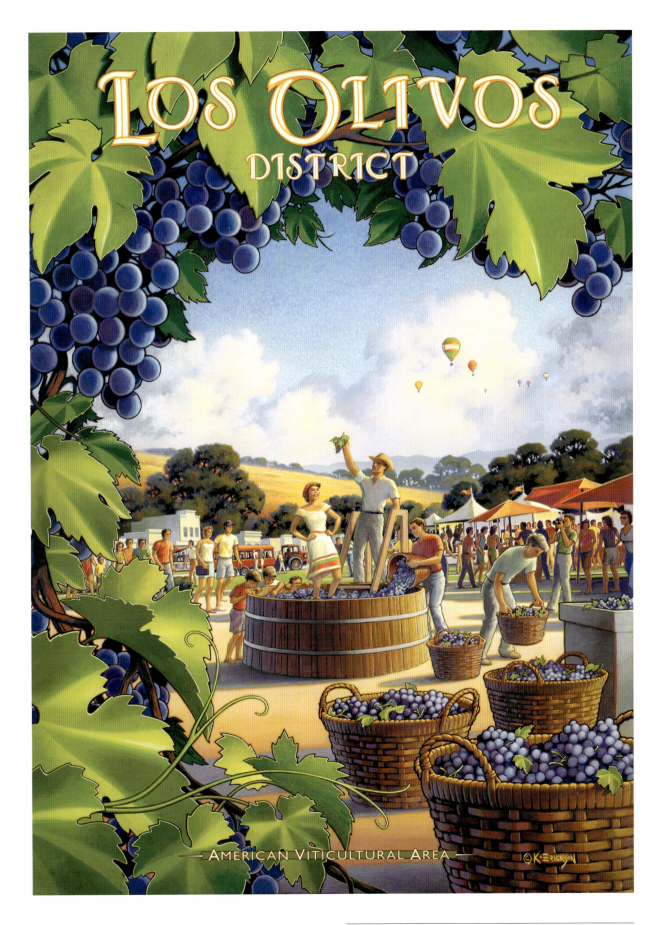

power. I imagine Einstein's shabby hair catching fire at the very thought of it.

Personally, I prefer living in the present and that day was no exception. While the sun rose and the crows crisscrossed the vineyards —chatty in the morning light — I felt more alive. Simply because it was the promise of a new day. The first light, briefly overcoming the surrounding mountain tops then washing over rows of the old vines, filled the canyon to the brim with warmth. The paired doves in flight, with their soft wings drumming, seemed to follow time as it moved along, as it has for eons, one sunrise at a time.

The old thriving vineyards have given thanks to this gentleman farmer. Beyond the sweat, the challenges, the consuming of his life, there was always a profound need to take care of something meaningful — a parallel in his own life. This one-night stand became a business endeavor lasting Bill to this day. And over time there had become only one way to approach it. That way is a perspective the Greenoughs refer to as the "Saucelito Way." As unique a perspective ever embraced by a family paying homage to the synergies of "place," it has become a continuum of tradition based upon family values.

Bill and Nancy's son Tom is a second-generation farmer and winemaker who embraces this "Saucelito Way." Bill is there to support and help guide his son when needed, while Nancy keeps the ship on course. Tom's own passion for the vines has overtaken his father's. I saw it firsthand during my stay.

Behind a colossal live oak tree on the ranch, whose strapping limbs once supported the wooden water tank used for the Ditmas' endeavors, was an earthen cellar. It was carved into the hillside beneath the oak tree's shadow and stored Ditmas' wines all those years ago. It was here that I met Tom. He spoke about the land with life experiences beyond his years. He spoke of the future, about planting the nearby brushy hills with vines. I could not help but be impressed by his vision. The young man who followed his father's path has himself caught the wine bug fever, with a passion all his own.

"It is a rare, orchestrated symphony of place and people producing world class wines, conducted by the Greenough family."

And Tom's efforts are bringing acclaimed wines to the table; their 1880 Zinfandel won numerous awards, along with their Petite Sirah and Arroyo Grande Valley red blends.

While we talked casually outside beneath the Saucelito Canyon sun, our attention once again turned to the oak. Tom mentioned an arborist had dated the tree to be close to 300 years old and said, "The oak tree spends 150 years of its life growing and another 150 years of its life dying." It was a profound statement: every life reaches a crescendo, a point when the ebbing of life begins to descend into the reflective pool of old age.

Yet here in Saucelito Canyon, where the old vineyards continue to thrive, time seems to have forgotten their presence. After over a hundred years, the ebbing of this vineyard's life has clearly not begun. Perhaps its lengthy life is due to what's buried within its treasured soils, bundled

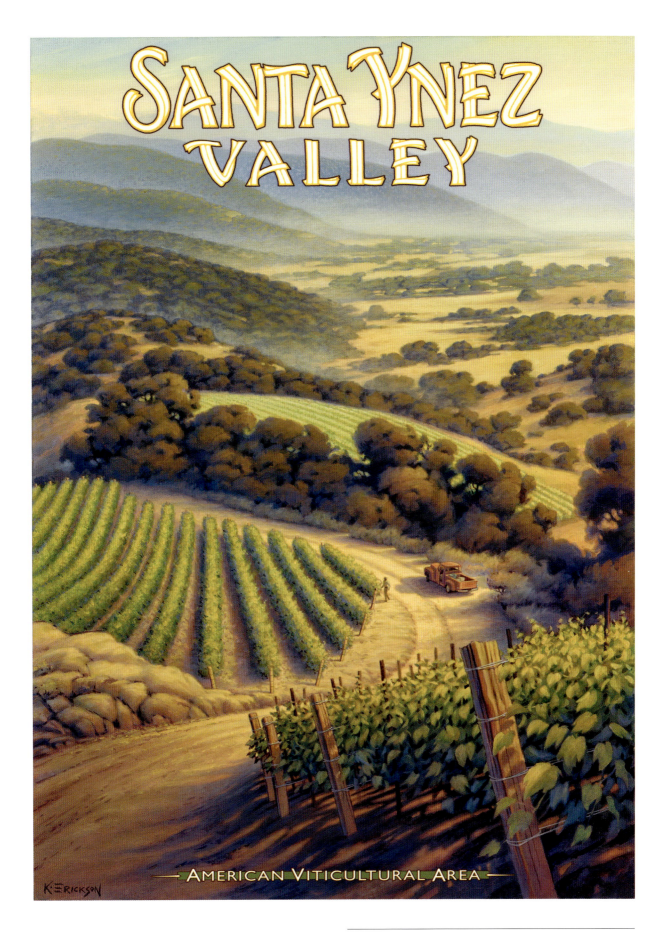

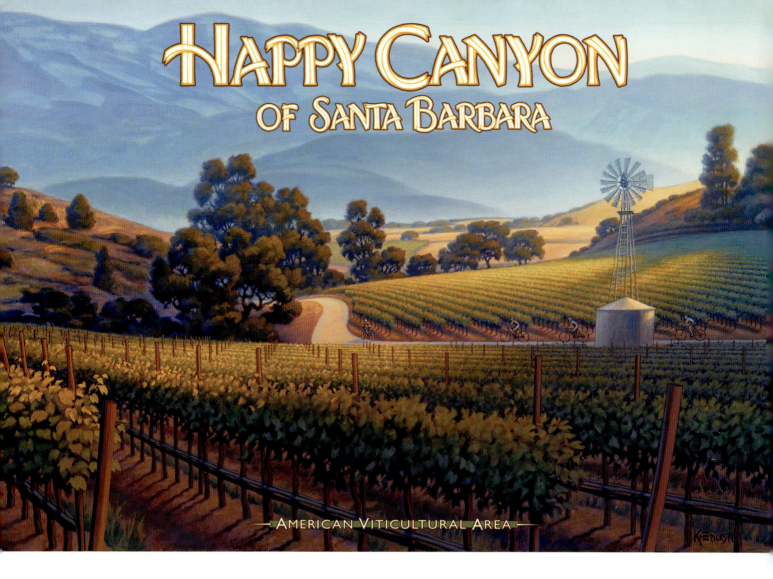

Happy Canyon
of Santa Barbara
— American Viticultural Area —

with the Greenoughs' "Saucelito Way," a fine balance sustaining life. In this remote place "off the grid," where a sense of timelessness continues to breathe over the land, these vines continue to celebrate life.

Leaving the Greenough ranch was like leaving an old friend. But there were other regions to visit. As I traveled south, the cooling Santa Maria Valley lay open to the west with its shining stars: wineries like Cambria, Bien Nacido, Byron, and others. Ahead was Solvang and Lompoc, two areas that have grown leaps and bounds in the past few years.

Passing the many vineyards and open grasslands, I soon came upon the quaint town of Los Olivos, where wineries like Fess Parker Winery and Firestone Vineyard adorn the town's fringes.

Before entering the heart of Los Olivos, I turned onto a country road which led me through a mountain corridor that traversed a short grade to the west. The road twisted in elevation until it opened up to Ballard Canyon. This was a place more reminiscent of Sonoma County — my hometown. A place where oak trees grew large and seasonal creeks babbled until late spring, turning thirsty and dry into summer.

Here, the canyons draw you in, welcoming you with hillsides couched in the most comfortable of settings. There is a shared communion between the acres of open grassland and vineyard presence. It's a pairing that complements both the region's natural beauty and its inescapable charm.

Santa Barbara County has a jewel here in Ballard Canyon, embracing each visitor with close to 8,000 planted acres and growing. It's a draw. The sub-appellation was established in 2013 and is relatively new. Its sand and limestone soils, inclusive of a canyon that runs north to south, is only a small measure of a larger vineyard equation. The Canyon is nestled between the cooler Sta. Rita Hills AVA to the west and the more heated Happy Canyon of Santa Barbara AVA to the east. This microclimate, with wind, sun and fog, is strictly Ballard Canyon's own. The vineyards here consist mostly of Syrah and Sangiovese, and smaller blocks of Merlot, Cabernet Franc, and Sauvignon Blanc.

Located in the sweet spot of the alluring Santa Ynez Valley there are a host of vineyards like Beckmen Vineyards and Larner, yet there is only one winery with a single tasting room, a gorgeous setting, and immaculately cared for vineyards. It is Rusack Vineyards and Winery. With 17 acres of planted grapes, attention to detail complements the allure of these low-lying hills. Exquisite rose bushes adorn the entrance to the winery. Rusack's promise to the visiting wine enthusiast is that taste really begins at the front door.

I had the pleasure of sitting with Rusack's winemaker, Steven Gerbac, beneath the canopy of

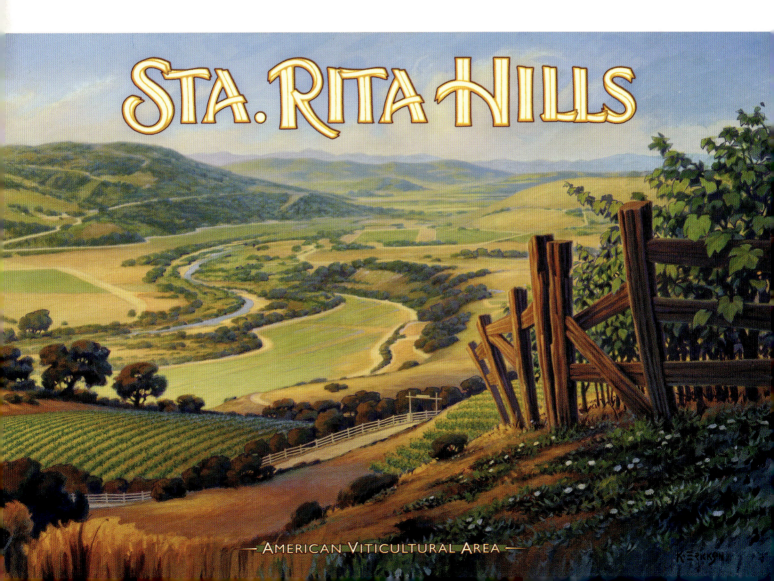

Sta. Rita Hills

— AMERICAN VITICULTURAL AREA —

a host of live oaks. It was outside the winery's tasting room, overlooking a block of vineyards. He and several others, including Wes Hagen, J. Wilkes' winemaker, championed the AVA designation. They drove the Ballard Canyon hilltops, defined the landmarks, and at the end of the day drew the plans and submitted them for federal designation.

Having first worked as assistant winemaker here, Steve's mastered the art of fine wine in the very appellation he helped define. He's putting his years of experience and firm belief of the region's potential to good use. Like most winemakers, he knows the vineyards, and moving through the ranks of the Rusack venture, knows them well.

Of course, having a tasting room attached to the winery staked profusely within its own vineyards, is a synergy that's working for both the Rusacks and Steve, along with throngs of wine enthusiasts seeking great taste. The wines, like the view, don't disappoint.

There is something surprisingly fresh about a winery whose estate has the ability to bottle and pour wines where their vineyards thrive. To taste the wines that come from the vineyard one is viewing is to experience the nuances brought to perfection in the glass. Its pairing is a presentation not found with off-site tasting.

The very essence enhancing wine is "place," and to be a part of that when drinking wine has relevance. Here, our senses breathe the same air, feel the same sun, and look upon the very soils the vineyards thrive in. It is getting to know the wines from the inside out.

It's a pairing that complements both the region's natural beauty and its inescapable charm.

I understand the need to market where tourism has its draw. Satellite tasting rooms are the new thing for many wineries. It's business and it's working. But when you step onto the grounds of a winery like Rusack — sit beneath the welcoming arms of shading oak trees while looking upon the vineyard whose wine you hold in your hand — there's a certain inalienable vibe to the experience. It's real and never disappointing.

Steven pointed towards vineyards once planted with grape varieties less appreciative of the Ballard Canyon climate. "The vineyards have been replaced by vines that are more conducive to the land," he says. It's become standard practice, one that every grower has embraced: plant what fits.

The following afternoon I met with Geoff Rusack. We sat beneath the same magnificent oaks overlooking the large crowds that once again had gathered to taste, laugh, and relax. As we spoke about Rusack Winery and his Santa Catalina Island endeavor, I noticed Geoff's eyes wandering. He wasn't impressed by his grand achievements within the industry. While attentive to our conversation, he appeared infatuated by the imbibing crowd. It was a telling moment. His subtle smile revealed what matters deeply to those who care about this business of wine. Beyond the financial goals or responsibilities, there comes a great sense of satisfaction derived by pleasing others. Geoff wasn't a conglomerate executive grinning about wine sales or stock prices. There were no "cha-chings" sounding off in the back of his head as he watched the wine crowd drink it in. Rusack was pleased because they were. He, like the many men and women working in the wine business, has a distinct love affair with people. After all, this is a business rooted in the world of hospitality.

And while Geoff was clearly humbled by the sight of the crowd, he was vicariously sharing in their celebration. The moment helped me better understand this complex business of wine and its human connection. The personality, character, and nuances that define the best of wine also define the best individuals, the men and women whose monumental contributions to the bottle go unseen.

5
The South Coast
American Viticultural Area

Comedian Fred Allen once said Southern California is "a fine place to live, if you happen to be an orange." That was then. Today it might be said the South Coast is a fine place to live if you happen to be a grape — or more fittingly, married among them.

This isn't due to the region morphing into a grape-growing mecca. It's about wineries scoring big with wedding venues, barrel room parties, and posh event centers. Wineries have become a happening place in the South Coast. They focus on the fact that profit margins are no longer simply based on drinkability, but also the successful marketing of these important events. With the region's proximity to a colossal population, it's a big "Oh, hell ya!" for every winery.

Yet before the population boom, concrete channels, toll roads, and the Kardashians, Southern California was a land defined by its agriculture. The warm climate influenced people to relocate in droves from across the country, settling among its mouth-watering citrus groves. It was a renaissance led by an aggressive turn-of-the-century travel campaign, marketing what the finest wines define best: sense of place. This was not only about living comfortably, but let's face it, about living fat.

The allure was undeniable. Agriculture was huge. Business opportunities were exploding. Aircraft makers like Douglas, Hughes, and Lockheed eventually headquartered here. Hollywood brought industry and celebrity to the state as it continued to grow; not just massive egos, but concrete blocks and plump, juicy fruit.

The South Coast continues to be a branded paradise. Even today's vineyards are finding some love on the outskirts of the thriving and towering steel metropolis. Scattered among the unincorporated hillsides and valleys, void of concrete structures, are 3,000 acres of planted vineyards. The region's heavy air, ocean influence, well-draining soils, and sun speak of a region not only home to today's massive infrastructure but one suited to vineyards. Varieties like Cabernet Sauvignon, Chardonnay, Zinfandel and Sauvignon Blanc thrive among their neighboring urban digs.

The South Coast AVA was established in 1985 and includes five counties: Los Angeles, Riverside, San Bernardino, San Diego, and Orange County. With an AVA population of approximately 20 million, it's colossal. Temecula Valley, Ramona Valley, and San Pasqual Valley are among South Coast's sub-appellations.

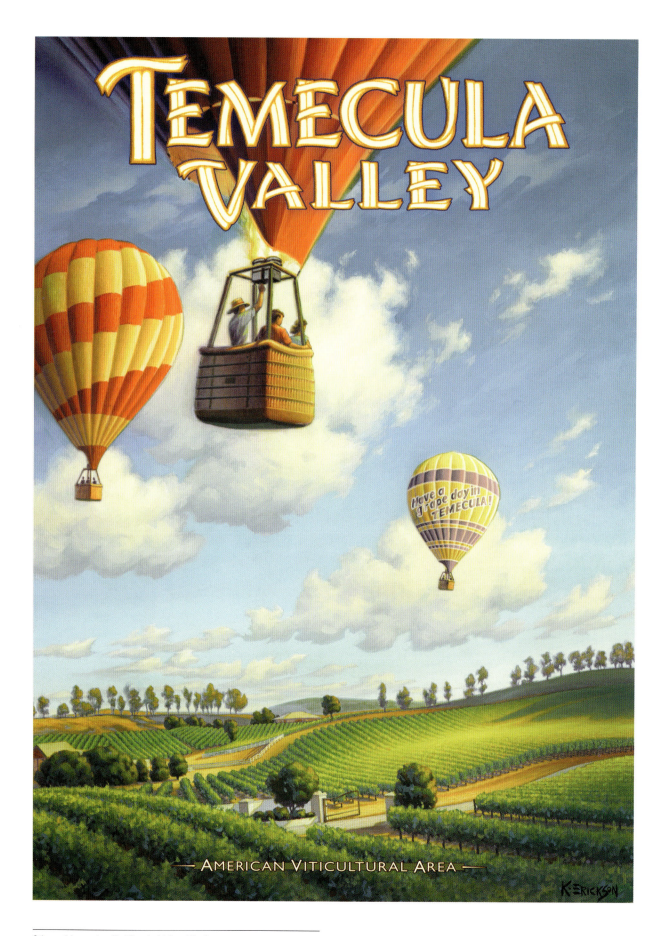

Along Temecula's Rancho California road, where Franciscan priests once traveled, is where sacramental wines of the mid-1700s brought the first vines to the region. The 1960s brought the boom, a renaissance that kickstarted Temecula Valley's flourishing grape-growing region.

Today, Temecula Valley is the region's largest wine producer, home to almost 1,300 acres of planted vineyards. Cooling breezes from the neighboring ocean and some serious heat from the east bring life to the region's signature varieties. The vineyards benefit from low coastal passes like Rainbow Gap and Temecula Gorge that open and reach inland. The heavy ocean air is drawn inland due to rising warm air from the east.

When you hear the name Temecula or "Temecunga," as the Luiseno Indian once eloquently referred to as "place of sun," there remains an indigenous romance to this sub-appellation. Today the region is not only about its wines; for many, it's about the road trip to the AVA's most undisputed wine country party draw. From Los Angeles, Riverside, San Diego, and beyond, the hipsters, baby boomers, and millennials alike are drawn to the region from their nearby digs. They come to indulge in the region's abundant play zone as much as they come to taste. Take Temecula's New Year's Eve Grape Drop that attracts a sizable crowd, or the hot air balloons lifting people high above the vineyards. There is also the Wild Women of the Wine Country 5K hosted by Wiens Family Cellars and Ponte Winery. Personally, I'm waiting for the first winery to sponsor sleepovers among the vineyards.

The first commercial winery was established in Temecula by golf icon Eli Callaway Jr., whose vineyard plantings began in 1969. The region has grown immensely since Callaway's first endeavor.

Due to his success, Temecula made headlines when Queen Elizabeth II and his Royal Highness partook in two glasses of Callaway's 1974 White Riesling at a luncheon in New York. The Queen requested to meet the vintner and gentleman farmer. Nothing like a hole in one for Ely Callaway Jr.'s early wine endeavor.

After Callaway sold the winery to conglomerate Hiram Walker & Sons in 1981, he went on to make another hole in one with the creation of Big Bertha, a monster-headed club that would become a golf industry sensation.

Although Temecula speaks of a '60s renaissance, it is the Cucamonga Valley in San Bernardino County that is recorded as having the greatest grape-growing history. Early endeavors began in 1838 with Tiburcio Tapia at his Cucamonga Rancho. Here is where the region's vineyards and dry farming has its roots. As the region grew, it was the early pioneer grower Secundo Filippi who boosted it to become the largest grape-growing region in the world, recorded to have planted more than 20,000 acres prior to prohibition. It was also Secundo Filippi who founded the Italian Vineyard Company (IVC) in 1883 and brought the first massive wine production to the Cucamonga Valley. The fourth generation of Filippi farmers have been instrumental in establishing the 1995 Cucamonga Valley AVA that today has approximately 1,000 planted acres. Unfortunately, infrastructure is gaining the most traction in the region. Its concrete, steel, and massive pavement is far more favorable than growing vineyards. The necessary four-lane freeways are a place where you get to know your neighbor's bumper well.

Beyond the land-gulping metropolis of Los Angeles' South Coast AVA is the newest AVA: Malibu Coast. Though not part of the South Coast AVA, it is a neighboring region beginning near the Pacific Coast Highway. Here, the few blocks of seaside vineyards begin their climb high above Malibu's extensive rooftops, reaching for the higher-elevation Newton Canyon and Saddle Rock sub-appellations.

In Malibu Coast, smaller vineyard blocks are most common. They adorn the dry, steepened landscape where growers like Gabrielle Harris of Sage Hill Vineyards is bringing acclaim to the region. Her 2013 Syrah-Cabernet Sauvignon blend won a Bronze and her 2010 Syrah made "special mention" in the prestigious 2015 International Decanter Awards. I had the pleasure of visiting with owner-grower Gabrielle Harris at her quaint vineyards rooted among the steep Santa Monica Hills. She was delighted to have "the first Malibu wine to win this international competition," out of 15,000 total entries.

She's not alone in her success. Today, Malibu Coast AVA has over 50 vineyards throughout the region, with others achieving recognition themselves. Many individuals with past business endeavors have moved onto grape growing. They are financially secure. They've come to know the value of life among vineyards and the quality of wines that can be produced here, as well as the challenges that come with them.

With present laws not permitting wineries to exist in the Malibu Coast region, growers have adopted assistance from co-op facilities like the nearby custom crush in Camarillo. Here, local growers produce their own wine with in-house winemakers who understand the craft.

At this time, there are 210 acres of vineyards planted in the Malibu appellation and nearly 50 commercial wine grape growers, with not one winery to call their own. Of those planted acres, varieties like Cabernet Sauvignon and Merlot grow in the higher elevations, while varieties like Pinot Noir and Chardonnay grow closer to the coastline. Syrah, Malbec, Sauvignon Blanc, and many other varieties also call Malibu their home; one- to seven-acres of planted vineyards is average.

Although vineyard plantings have descended upon this peaceful land — and are settling in beautifully — obstacles and challenges threaten their future.

One significant challenge confronting growers is a new era of prohibition threatening the Malibu Coast. It's not about banning wine production as in the past — an era that leveled the industry — but about prohibition with vineyard plantings. That's right: it's about putting a halt to vineyards, the very attraction both travelers and wine enthusiasts appreciate. Limiting tasting rooms in the Malibu County hills, the Land Use Plan (LUP) restrictions hamper public access to some of Malibu's most charming hillside vistas, naturally beautiful places to relax at the table while eating and drinking. Fortunately, you can thank grape growers like John Gooden who continue to champion the Malibu Coast AVA.

I visited John and his wife DeDe at their Montage Vineyard and home in Malibu. John pointed to the outdoor table where the makers of the Malibu Coast AVA had sat and drafted the application. "It all began right there, at that table," he said admiringly. If you understood the paramount efforts involved in getting an AVA designation, you would understand the time spent at that table: drafting notes, conducting historical research, and finding words to support the region's beautiful terroir. It was more than a challenge, more than financially draining; it was daunting. Fortunately, the growers are doers. They understand markets. They appreciate taste. They were passionate about bringing the AVA to fruition because they knew the quality of the grapes born from this unique land was exceptional.

We walked through the Gooden's small terraced block of vineyards below their house. It was late February and bud break was happening. DeDe, who is also a partner in the endeavor said, "We had a good rain that was followed by heat," which attributed to the spring-like burst of green bud and leafing.

"Vines need a consistent period of cold weather in the mid-50s to rest," said John. The region is noted for its short dormancy but if bud break comes too early, it may lead to a host of serious issues, mainly frost that can attack the frail bud and destroy an entire season's yield.

Later in the day, we took a drive to a neighboring block of vineyards. It is a magnificent estate that John Gooden helps manage, and not for any reason but because he wants to. It's also within Malibu city limits. We stood on a steepened hillside beside a block of Chardonnay, overlooking the ocean in the shadow of a healthy sycamore tree. As John reflected on the bud-breaking vines he brought to mind what is commonly valued here. "We're family-owned," he said. "Everyone."

His words resonated with the essence of why wine matters. It's about more than just drinkability. It's about relationships. Growing grapes is as much a family affair with growers as cooking is in the kitchen. It's about being together and helping one another. It's about making something special.

> **They were passionate about bringing the AVA to fruition because they knew the quality of grapes born from this unique land was exceptional.**

While Malibu may be perceived by many as dripping with money and having buttery notes of coastal bliss, there's more to this legendary "endless summer" charm than a place to spend and surf. Its resources are reflected in the wine and in the uniqueness of the land, defined by a federally approved American Viticultural Area. Malibu Coast also has its larger wine growers like Cielo Farms and Semler Malibu Estate Vineyards, who have led the charge in the region. Semler's ranch, once an avocado farm in the '70s, was lost in a ravaging frost. Today there are vineyards, a zoo safari, and with good fortune, a winery in the making.

I met with Dakota Semler at the tasting room. He took me on a drive through the Malibu Estate Vineyards. Rather than traveling among the vineyards in the customary ranch truck, we traversed the property in a zebra-striped Tesla. It wasn't your ordinary sedan. Of course, this wasn't your ordinary vineyard either. When we entered the heart of the vineyards there were barking zebras, water buffalo, and Stanley the giraffe from the movie *Hangover 3*. It seemed a fitting ride. Technology has come to the wineries. Whether by iPad, cellar innovations, weather alarms, or in this case, a custom Tesla, tech gadgetry is evolving no less than the wines themselves have over the years.

To visit Semler is to know something's happening. It's more than technology, more than the adventure. The Semlers provide a visual wine tour like few others, and with 2,000 visitors per weekend indulging in its safari tour and open-air tasting room, it has become a hip way to indulge. But it's not only about Stanley the giraffe peering over the fence posts; it's about the wine, quality grapes, and varieties that thrive in a land that's producing high-scoring wines.

Dakota pulled alongside the vineyard road. With the sound of a "hee-hawing" zebra in the distance, we stepped out for a short walk with a view of the fertile valley opening beneath Turtle and Saddle Rock. This was a land rich with Chumash tribe legend and noted history. Within Turtle Rock's caves, a rock's throw away from the vineyards, are petroglyphs, an indigenous bit of Native American history.

I asked Dakota what led him to this wine country endeavor. He told me that prior to stepping knee-deep into the wine industry, he had thoughts of becoming a commercial underwriter. There was good money in it. But "on a wine trip to the Mediterranean," he said his father took him aside and asked, "Do you want to work a hundred hours a week at a job doing what you don't love? Or work in the vineyards that you do?" It was a seminal moment for Dakota, whose passion for carrying on the legacy is today both intimate and revealing, and he has never looked back. He speaks of the vines that are thriving. He talks of the history and the joy of being a part of the vineyard, this sense of "everywhereness." It was remarkable, not because of the buzz associated with the industry, but of the history and the continuum of family.

To understand Dakota Semler's decision is to know he is a part of something larger than himself. He is part of a family that began its journey with avocado farming and is now host to an animal sanctuary and a vineyard with acclaim.

Even Stanley the giraffe is feeling the love.

Another legacy in the works is Richard Hirsh's Cielo Farms. Here the sense of place resonates from within. The stone taken from his property was used in the construction of the tasting room walls.

I had the pleasure of tasting a delicious Woodstock-labeled Cabernet Sauvignon-Merlot blend with Richard Hirsh and John Gooden. We sat at a lengthy rustic table within the tasting room's heavy stone walls.

Richard explained how the Woodstock label reflects the essence of the iconic music festival at Max Yasgur's 600-acre dairy farm — a "cultural touchstone," one that crossed borders of race and the establishment. It was a moment of monumental peace, music, and togetherness. The Cielo label is a tribute to that moment, reflective of the open love the era spoke of, and what wine itself reflects.

While tasting, John said reverently, "I remember the day I was at Woodstock." This experience was a touchstone in his own life. It wasn't surprising. I think we all have been "there" in one way or another. John was one of the fortunate to have lived it.

Richard has created his own place here on the mountain top above the Malibu Coast, both breathtaking and reflective of the peace, beauty and spiritual culture his label represents. The costs involved with building a winery and tasting room of this significance can be monumental. Yet, most of these owner/growers have made their dough. They understand the financial callouses associated with vineyard farming are immense, that weather worries and regional dormancy issues will knowingly hound them. But "rolling up your sleeves" is in their blood. It's a community with a history of wealth born by fortitude that is neither flaunted or abused, but healthy and sustainable. Ironically, while the growers' lifestyles smack of good living, it was pointed out to me by marketing professional Dan Fredman, who works with the estate, that "there are more rehab centers allowed in Malibu than tasting rooms."

It's no revelation that celebrities and those with financial wealth have chosen Malibu as home, or, for those struggling, as a place to recover. They are passionate about healthier living and creating a better lifestyle.

For the grape growers in Malibu, it's about less regulation. This allows them to plant vineyards that may not be indigenous to the land, but do speak of man's communion with it. This in of itself is a beautiful thing, a matter of relevance. Scattered about are the native pines, sycamore, and scrub oak. The buck brush, chemise, and choking thickets bask in the ocean's reflective light and heat, soothed by maritime breezes. The vineyards themselves adorn this indigenous landscape with stitching of green in summer and patterned colors in fall.

This natural beauty is captured at different twists and turns as you drive Mulholland, Kanan, Point

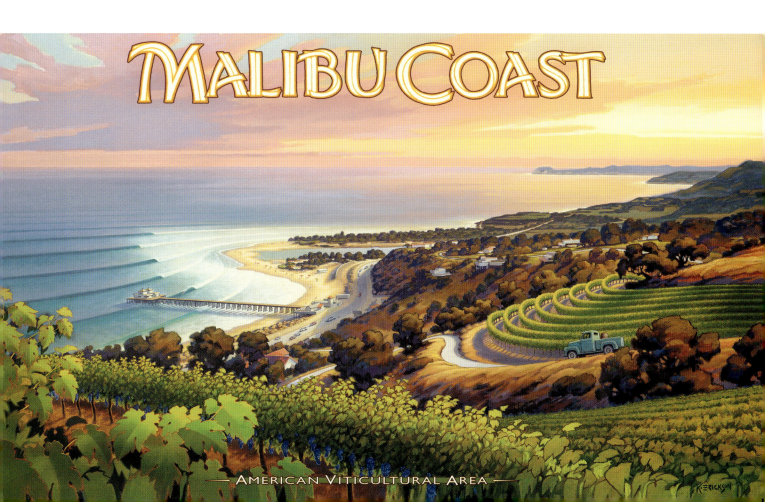

MALIBU COAST
— AMERICAN VITICULTURAL AREA —

Dume, Latigo Canyon, and along the Pacific coastline. The turning roads lead you in. They take you down and lift you up. Like elegant drapery, vineyards adorn the harshness of their surroundings, gently humanizing the region's aesthetics.

These vineyards are rooted in the signature terroir, which is comprised of climate, minerals, elevation, the rain that falls, and drought years that follow. It is what makes Malibu wine its very own. Each region embodies the essence of its land's personality. What other drink speaks so reverently of the land's natural indigenous character than wine?

> **What other drink speaks so reverently of the land's natural indigenous character than wine?**

Not too distant from the Malibu Coast, beyond the curling set of waves lapping at its legendary shores, is California's most celebrated South Coast island, Catalina.

Catalina Island is located approximately 20-something miles across the gentle and tumultuous blue Pacific waters. Although not a designated AVA, the island's vineyard presence has a visceral sense of place unlike any other in the region — or throughout the state — for that matter.

Here, Geoff Rusack and his wife Alison Wrigley Rusack (an heir to Wrigley Gum fortune), owners of Rusack Vineyards, are daring to dream big. Their five acres of island-planted vineyards may appear small, but the challenges are many.

I had the privilege of touring the island with Geoff. We left the quaint harbor town of Avalon on a clear sunny morning to visit their 85-year-old El Rancho Escondido, a magnificent Spanish-style mountain retreat and former iconic horse ranch. Of course, this visit wasn't about equestrian matters. It was about viticulture; specifically, the idea of transforming an iconic 640-acre Arabian horse ranch into a tasting room and world-class winery. This is the first and only commercial wine venture on a California island.

It was a spectacular morning. The visibility was as clear as the rich history of El Rancho Escondido, which goes back to a time when Alison's grandparents, Helen and Philip Wrigley, bred world-class Arabians on their Catalina ranch.

Today, the transformation from horses to vineyards is a natural fit. It is also an extension born by a stable of delicious wines they produce at their iconic Ballard Canyon vineyards and winery in the Central Coast. Only here on Catalina Island, the vines are more vulnerable as the harsh maritime elements come at the vineyards from all sides.

This hasn't deterred the Rusacks. Determined to grow island grapes and produce fine wines, they planted five acres of vineyards in 2008 at their El Rancho Escondido with Chardonnay, Pinot Noir, and Zinfandel.

Zinfandel is at the heart of the family's most recent story. When Geoff heard there were Zinfandel vines on neighboring Santa Cruz Island in the mid-1800s, it piqued an interest. He and his three sons, Austin and Parker, hiked the rugged terrain of Santa Cruz Island in search of the creeping vines. Successful in their efforts, they had the clippings propagated, and by doing so, established a continued history of Channel Island plantings.

On Catalina Island, the vines are more vulnerable as the harsh maritime elements are coming at the vineyards from all sides.

Rusack Vineyards' Central Coast viticulturist Larry Finkle believes the island has influences similar to the Russian River Valley appellation in Sonoma County, yet there are many noticeable differences between the two distinct regions. One such difference is the treatment of deer fencing. While deer fencing is installed throughout many north coast regions, for these island vineyards it's a whole other story. Larry has installed two- and three-inch steel posts — as opposed to the more common T-posts growers often use — to anchor the Rusack vineyard fencing. Sound like overkill? Sure, if not for the beastly bison grazing the open hog backs!

Brought over in 1924 for *Vanishing America* — a movie based on Zane Grey's novel —the bison now freely roam the island. They are as fond of grapes as any creature with an appetite for the fruit — although these represent a cud-chewing population in the state that is bigger than most.

In addition to bison and deer, Catalina Island was once home to wild horses, grazing goats, and boars, all of which were eventually removed from the island due to their non-indigenous presence.

Foxes and other native animals continue to inhabit the island, but the bison and deer have yet to be removed. There are those who feel they play an important role in the cultural fabric of the island. Maybe that's code for tourist attraction. Or maybe it's the Hollywood in the bison. Let's face it, they are celebrities and no one wants to knock a star, especially a heavyweight with a big head and sharp horns.

Unfortunately, I never did see a bison while on my travels. It wasn't really a bother, there was so much more to see, especially with the day being strikingly clear and the mainland looking so remarkably close. "It is rare to see so far," said Geoff. In fact, on his flight over to the island's "Airport in the Sky," it was the first time he could see all of the Channel Islands from the air at once.

Geoff has seen a lot from the air. As a former aviation attorney, his appreciation for air transport and planes is also very clear. While having coffee at the airport, he pointed out his sleek riveted Cessna Caravan 10-seater. He explained the large pod beneath its pregnant-looking underbelly was used to transport his cases of Catalina Island wine back to the island, wine from island-sourced grapes that had been pressed, produced, and bottled at the Rusacks' Central Coast winery. A cargo Cessna Caravan provides care for the freshly picked fruit flown from the island to the mainland winery.

After leaving the airport, we traveled further inland and soon arrived at the Arabian horse stables, where the future tasting room and winery were presently under construction. Nearby were small blocks of dormant Pinot Noir, Chardonnay, and his adventurously sought-after Zinfandel vineyards, now managed by Larry.

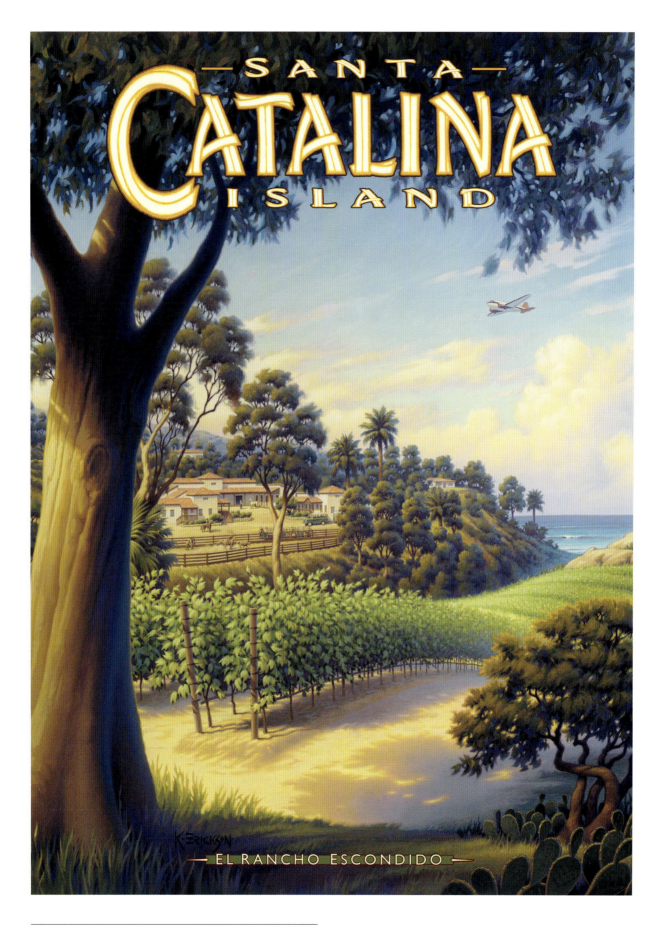

When speaking with Larry, it was clear the challenges facing island grape growing are tremendous. "There are wasps, dust mildew, high magnesium, and salt in the soil," he said. Another element that can decimate a crop during hang time are the Santa Ana winds. The winds that originate from the mainland can blow north of 50 miles per hour. They are heated and intense, and when funneling from the east, gain speed and power before blowing through the island with ferocity. Geoff says these winds create shatter— where grapes are forcibly pressed against each other until they burst —and at times may wreak havoc on a harvest.

I've anchored my own boat in Avalon harbor and have been tossed around by a severe Santa Ana windstorm. The sight of the crashing waves' foam kissing the tops of the highest palm trees is stomach-turning. Seagulls mostly don't fly; those which do, flop. Boats don't sail and moorings occasionally break loose. The potential damage these ferocious winds might do to an October harvest is frightening — and they blow in full force at the peak of harvest.

This hasn't swayed the Rusacks. They welcome challenge. With a built-in market population of 20 million people across the San Pedro Channel, there appears to be no stopping what they can achieve. When I asked Geoff if he was going to expand his vineyards he said, "I have no plans to grow more." I sensed this could change. The Rusacks are "allowed a total of 18 acres to plant," he said There is much more to do.

As we returned to Avalon I better understood what was going on with Geoff and Alison's island endeavor. It was a continuum, albeit one less associated with Chicago chewing gum, swinging baseball bats, or Arabian horses galloping across the Catalina headlands. Their journey was about reaching further and offering a distinct character of their own through their boutique stable of wines. Their premium wines speak not only of a place out in the sea, but of an iconic family's roots growing deeper and another generation tapping into the promise of something more than a continuing legacy or grand celebration of life, but the enhancing of its pleasures.